Peter an

Wil
the Mediterranean

Consultant: Nicholas Turland

JOHN BEAUFOY PUBLISHING

Wild Flowers of the Mediterranean

Introduction

This pocket guide, arranged according to flower colour and simple floral characteristics, aims to provide a key to the flora of the Mediterranean. Increasing numbers of nature lovers use their holidays for recreation in the sunny south where they encounter a rich and very different plant world. This volume with its handy format will fit into every jacket pocket or rucksack. It describes around 530 of the plants growing around the Mediterranean with illustrations of 286 of them, mostly in large format colour photos in their natural habitat. While this can only be a selection of the 24,000 plus species growing in the countries of the Mediterranean, the reader will find within these pages the most widespread, abundant and striking plants of the coastal vegetation, the evergreen shrub communities and the woodlands. Also included are the species of cultivated land, roadsides and waste ground in inhabited areas. The 47 most conspicuous ornamental plants are illustrated on the cover flaps. Only a small number of important trees and shrubs could be included from the deciduous, sub-Mediterranean zone of the adjacent mountainous regions, but none of the many species of the Mediterranean montane vegetation. In addition, only a very few examples of the numerous "endemics", i.e. species that occur in very small areas, could be included. We hope that this guide can provide initial assistance to all Mediterranean travellers with a love of nature in getting to know this interesting flora.

The most favourable season for this is without doubt spring, from March to May when most of the species are in flower. But the other seasons also have their appeal: various members of the coastal vegetation and those of damp sites flower in the dry hot summer, as do plants in the mountains. Autumn in the Mediterranean, when the rains set in, signals the start of a new vegetation phase and therefore the flowering time of some plants.

In many instances inset photos of a characteristic detail have been added to the main image at the top right. An additional species has often been illustrated in the left hand margin.

The scientific names follow the most up-to-date available sources in line with new molecular genetic research in some cases. The English common names mosty follow the database of the Botanical Society of Britain and Ireland (BSBI). Where there are no equivalent common names, just the scientific name is given.

We wish to thank the publisher and particularly our editors Claudia Salata and Carsten Vetter for their expert and friendly support.

Peter and Ingrid Schönfelder

Guidelines for using this book

There is a simple identification key in the contents (see p. 5) which will quickly lead to a series of pages from where, by leafing through and making comparisons, the relevant species or at least a similar one can be found. In most cases the initial classification according to the colour of the petals as white, yellow, red or brown, blue and finally green or inconspicuous flowers will be successful. However, as there are always transitions in nature, in cases of doubt you will need to check out the next most similar flower colour. This applies in particular to violet flowers, whose colours sometimes shade more towards red and sometimes more towards blue. Some flowers also have the peculiarity of changing their colour between opening and finally fading, e.g. from red to blue. Most of the orchids, particularly most species of Ophrys, have been classified under red or brown, even if they have markings of green, yellow or blue streaks or flecks. In the first four colour groups the plants are then arranged according to simple floral characteristics which can be found as a symbol at the top left of the page:

Flowers radial, maximum 4 petals or lobes: this covers primarily the Cruciferae, but also individual species with fewer than 4 petals.

Flowers radial, 5 petals, occasionally these are deeply divided into 2 lobes, in others they are fused into a tube with 5 lobes.

Flowers radial, 6 or more petals or lobes: these are mostly the 6-parted members of the monocotyledons, but also a few 6-parted and multiple-parted members of other families.

Flowers in heads: several flowers are united in heads which often appear to both non-botanists and pollinating insects as a uniform flower (particularly in the Compositae).

Flowers with bilateral symmetry with free (e.g. Leguminosae) or united petals, often with one or more clearly developed lips (Labiatae, Orchidaceae).

The plants with green or inconspicuous flowers have been divided into

Ferns, when young always with curled-up leaves,

Conifers: trees or shrubs with needles or scale-like leaves,

Deciduous trees, predominantly woody species growing in the form of trees and shrubs,

Herbaceous plants, these are divided into dicotyledons and monocotyledons, the latter mainly comprising a series of grasses and grass-like plants.

The text accompanying each photograph starts with the common and scientific names, whose nomenclature derives from the most recent information in various international databases, particularly the Euro+Med PlantBase and Med-Checklist (Bibliography, see p. 304), and with the name of the plant family followed by the flowering period.

A brief section in bold points out any special features of the plant, helps with identification or explains the name. It also describes the plant's importance for us, e. g. as a medicinal, ornamental or crop plant.

The **Identification** section first indicates the plant's height, then gives a description of the species, beginning with the habit and the arrangement of the shoots, then the leaves and flowers and their parts, and the fruits. If two dimensions are given for an organ, these refer first to the length and then the width. Additional features are described in the photo captions on the righthand page.

Under **Habitat** there is a brief list of the most important sites for the plant's distribution: in addition to the biotopes familiar in more northern parts of Europe, these are primarily evergreen forests, especially of oaks but also pines; and the shrubby habitats which have arisen from thousands of years of use – clearance, grazing, overgrazing – the maquis (2–5m high, dense scrub) and garigue (lower, mostly open and patchy scrub), including the eastern Mediterranean phrygana.

The distribution of the main species in the Mediterranean region and on the Canaries is shown on a thumbnail **map** in the lefthand margin beside the species name.

The map has a dot for the presence of a species in the following sub-regions: · Portugal · Spain · Balearics · France · Corsica · Sardinia · Italy · Sicily · states in the northern Balkan Peninsula · Greece · Aegean · Crete · Bulgaria · Turkey · Cyprus · Lebanon and Syria · Israel and Jordan · Egypt · Libya · Tunisia · Algeria · Morocco and · the Canary Islands.

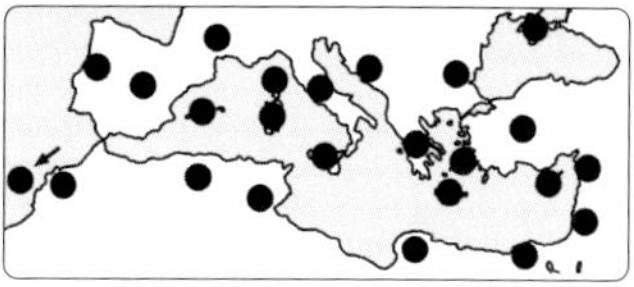

Under **Additional species** a description is given of an often similar and related species with the most important distinguishing features, a photo in the lefthand margin and its distribution. Details of flowering time, size and habitat are only given for the additional species if they differ significantly from the main one.

The most important **ornamental plants** are illustrated on the cover flaps and the first and last pages of the book, again arranged according to the flower colours white, yellow, red and blue (there are hardly any ornamental plants with green or inconspicuous flowers). These species, mostly with striking and decorative flowers, are found planted along streets, in public spaces and particularly in hotel grounds. Only individual species which have often become naturalised in the wild, such as the Hottentot Fig, have been included in the main part of the book.

Identification example

The shrubby plant shown here flowers in the South of France in May. The Table of contents leads to pages 158–169 following "Flowers red or brown" and "Flower radial, 5 petals". Leafing through the pages in question it appears that most flowers differ considerably in form and size. But there is an obvious similarity to the *Cistus species* on p. 164. The paler flower colour suggests the "Grey-leaved Cistus, *Cistus albidus*" which is confirmed by the description of the leaf characteristics: partially clasping the stem, with white felty hairs on both sides and 3 parallel, clearly raised veins on the underside. In contrast, the "Cretan Cistus" also illustrated has green or grey-green, 5–15mm long, stalked leaves. The distribution map shows that the species is known from France. Of the Cistus described under "Other species", one occurs only in the eastern Mediterranean and North Africa, the additional species *Cistus crispus* mentioned under *Cistus albidus* can be distinguished by more deeply coloured, very short-stalked flowers and leaves with wavy edges. This confirms the identification "Grey-leaved Cistus".

Grey-leaved Cistus *Cistus albidus*, South of France

The most important plant communities

All the species mentioned here as examples are also to be found in the identification section.

The **beach vegetation** is what the holidaymaker usually encounters first, as long as it has not been ousted by the important beach facilities such as rows of deckchairs and beach bars, which is often the case in the main tourist locations. But not far away along the beach you can often find the sequence of dunes with their typical species: the first land plants are already in evidence along the drift line, the level that the water reaches at normal high tides and where it deposits flotsam and nutrients. The low foredunes hold characteristic species such as the Prickly Saltwort and Sand Couch, the main dunes are usually covered with Marram Grass. In addition, from spring right through to the middle of summer, numerous interesting and sometimes thorny species grow here, such as the Sea Bindweed, Sea Daffodil and Sea Holly. The fixed dunes support shrubs such as the large-fruited subspecies of the Prickly Juniper. On sand further landwards the typical vegetation is pine forests. On the rocky coasts are to be found species which can tolerate the salt content of the spray, such as Rock Samphire and a large number of species of Sea Lavender.

The land around the Mediterranean would naturally be covered in **evergreen forests** at the lower levels, but these have been felled over large areas since ancient times to create agricultural land, for example extensive evergreen olive groves and also pasture. Nowadays you generally have to go in search of these kinds of woodlands. Several species of pine form stands on nutrient-poor soils, such as the Stone Pine and Aleppo Pine, while on richer soils it is primarily evergreen oaks

Dune vegetation with Sea Holly, Naxos

Holm Oak forest, Mount Lepini, Italy

such as the most widely distributed Holm Oak.

At higher altitudes, in the sub-Mediterranean or montane zone, the vegetation changes to **deciduous woods** in which, besides species of oak, there are also other deciduous trees and shrubs such as the Manna Ash, Hop Hornbeam and Oriental Hornbeam. In contrast to the Holm Oak forests, which are usually dark, these often have diverse herbaceous undergrowth.

After forests are cleared it opens the way for light-demanding **evergreen shrub** communities, which are much more species-rich than the woodlands. Depending on its height and composition, dense scrub growing to 2–5m in height with, for example, Tree-heath and Strawberry Tree, is known as **maquis** or macchia. Low scrub with locally differing dominant species is known generally as **garigue**. However, depending on the dominant shrub species, there are regional formations which are

Aleppo Pine wood, Italy

Garigue with Shrubby Kidney Vetch and Cat-thyme Germander, Naxos

Maquis with Tree Heath and Strawberry Tree, Corsica

named accordingly, for example the **tomillares** (with stand-forming Common Thyme) in the western Mediterranean or the **phrygana** with very varied species composition such as *Phlomis* species or Small-flowered Cistus in the eastern Mediterranean. In the gaps between the shrubs the interested botanist will find numerous annuals, especially in spring, but also bulbous and tuberous plants, including many orchids. In order to maintain this diversity of herbs for grazing sheep and goats, these stands are repeatedly cleared by burning. They would otherwise revert to the original woodlands over the course of centuries. Dwarf shrubs become rarer in the driest areas and grasses can dominate, forming steppe-like vegetation.

On **cliffs** and in **crevices** and **gorges**, in addition to a few widely distributed species of fern, endemics – those plants which are specific to

Old Olive trees with Bermuda Buttercup, Mount Gargano, Italy

small areas – have often survived. These have mostly been omitted from this book due to lack of space.

Extensive areas of **farmland** in the Mediterranean are characterised by the evergreen olive groves, which originally had a species-rich undergrowth but are nowadays often treated with herbicides and irrigated, making them botanically less diverse. In some areas citrus-growing is possible due to intensive irrigation. Viticulture and cereal-growing are now carried out in the Mediterranean, as further north in Europe, with the heavy use of pesticides, so that the characteristic wild herbs and undergrowth have become rarer or reduced to marginal areas.

With a generally frost-free climate, subtropical ornamental plants can be grown in **towns and villages**. With their large flowers these often catch the attention of the traveller before the mostly inconspicuous members of the Mediterranean flora. Many species, however, depend on irrigation. The most important of this group of plants are shown on the cover flaps.

Viticulture, southern France

Botanical terminology

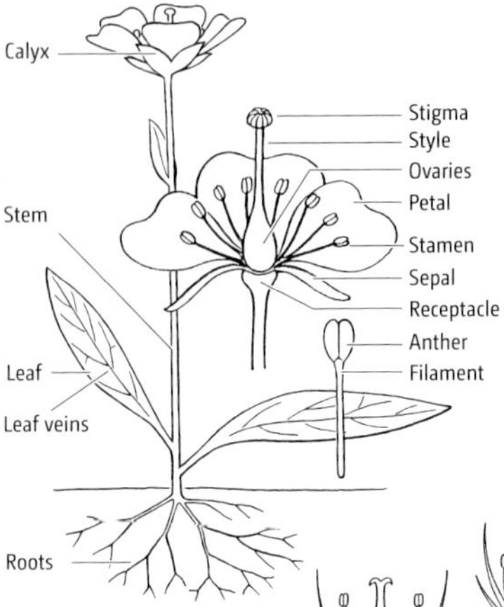

Shoot arrangement

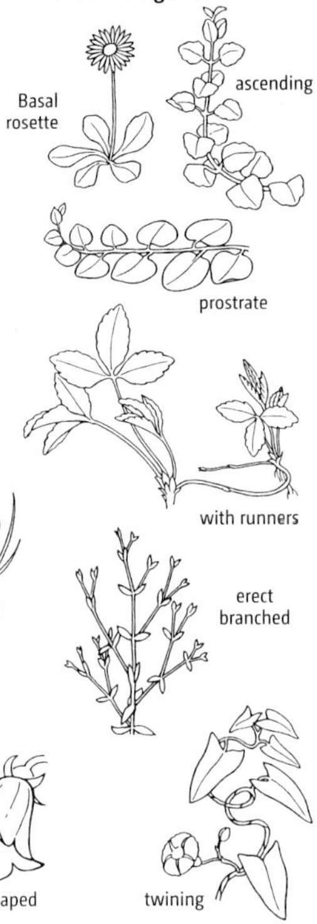

Ovaries

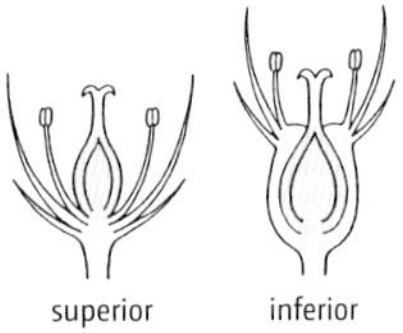

Corolla radial

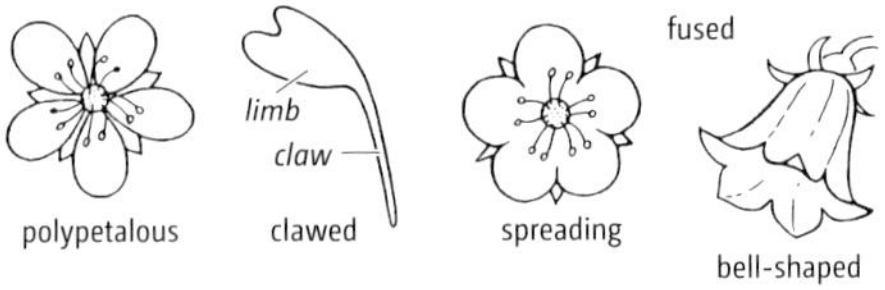

Corolla bilaterally symmetrical

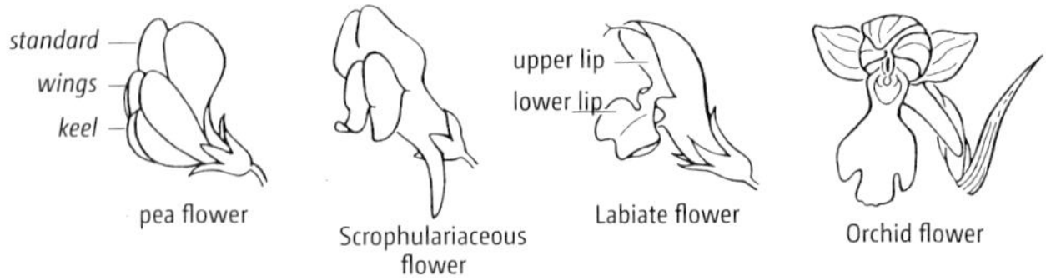

Calyx

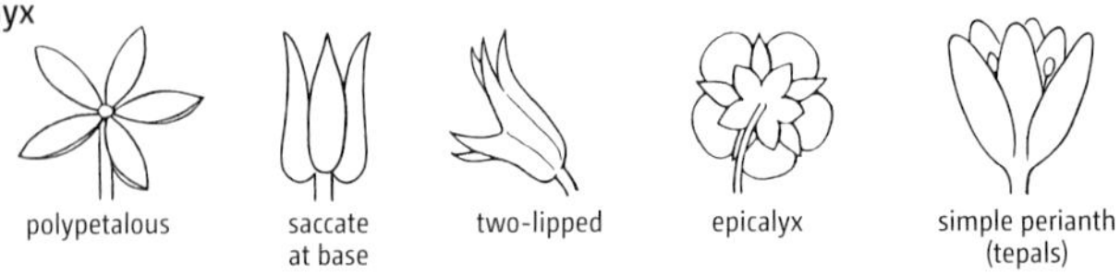

Inflorescences

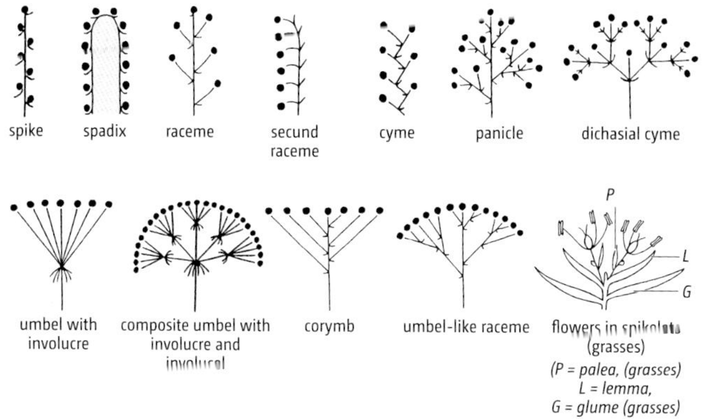

Flowers in heads

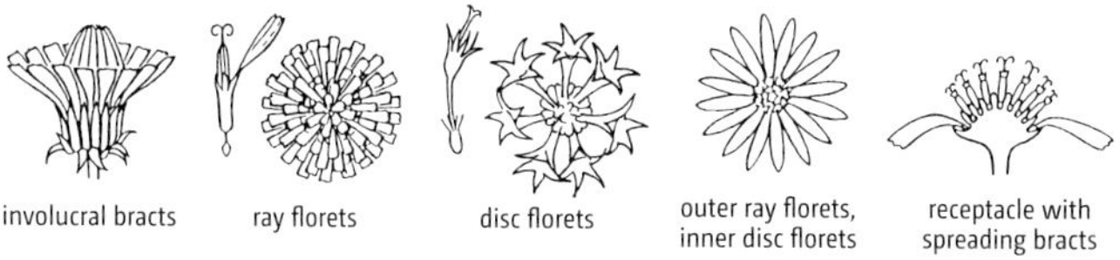

schizocarp

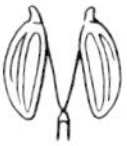

indehiscent fruit

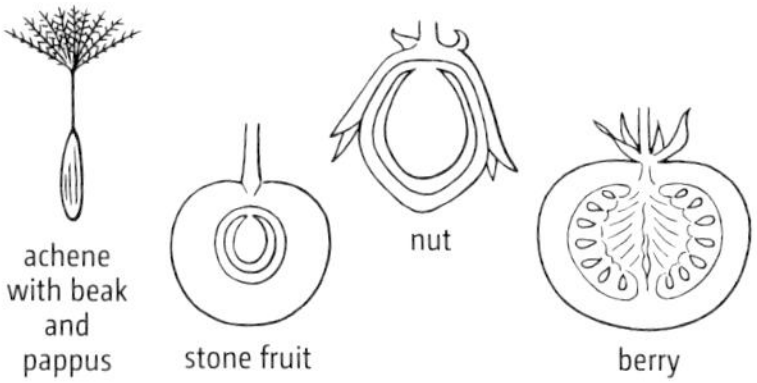

dehiscent fruits

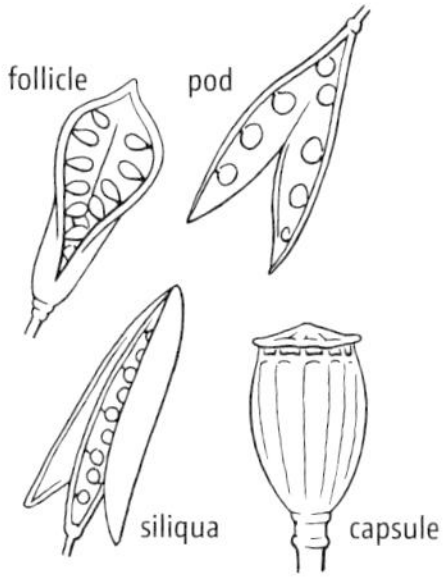

Aggregate fruit

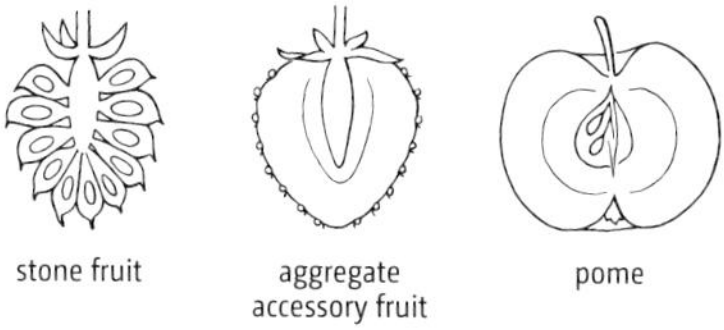

The leaf

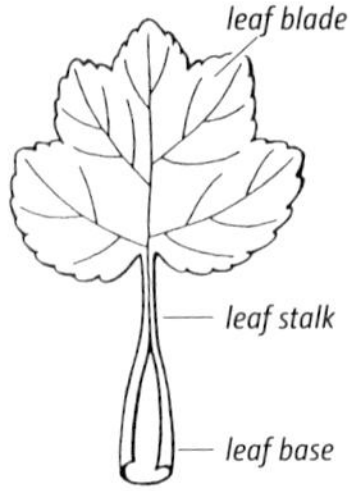

Leaf blade

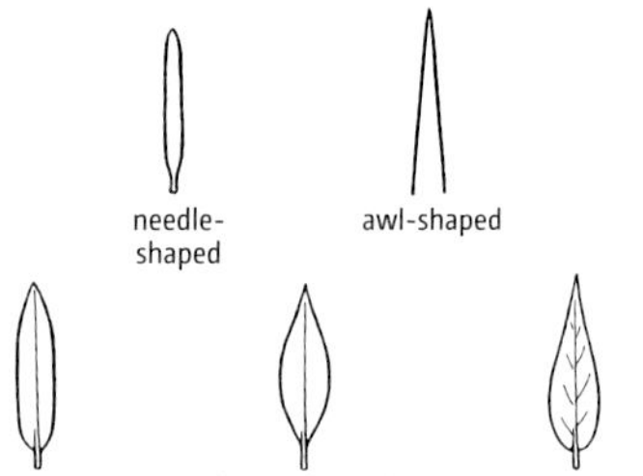

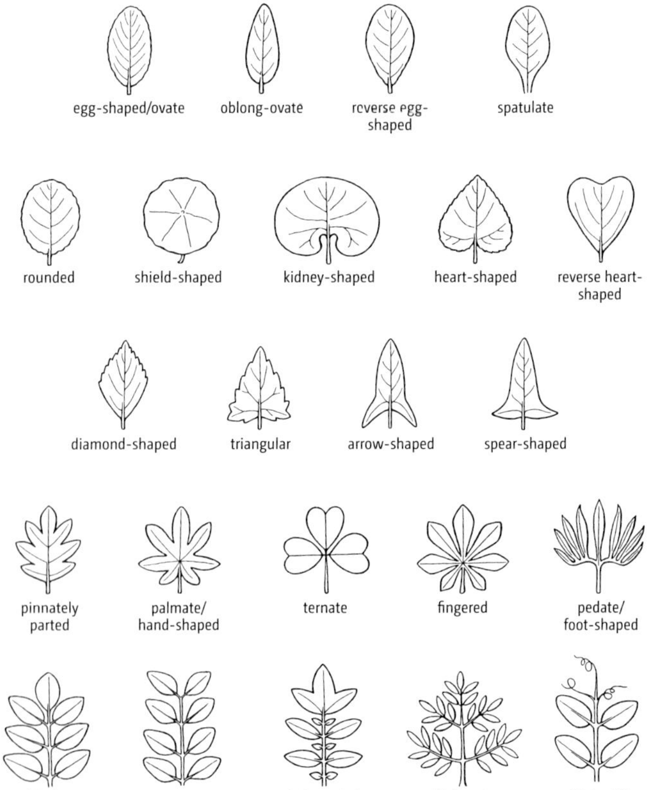

Leaf margin

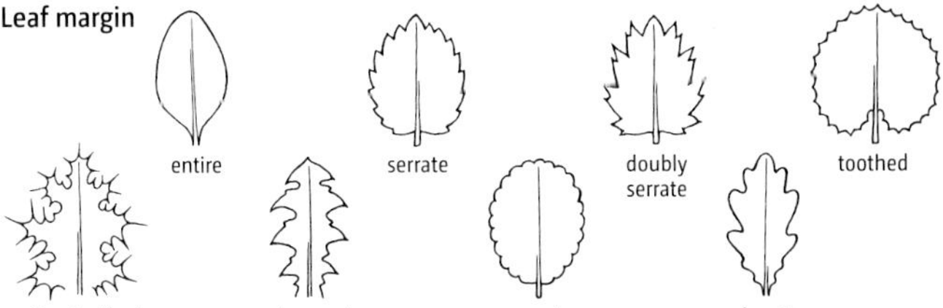

Leaf attachment

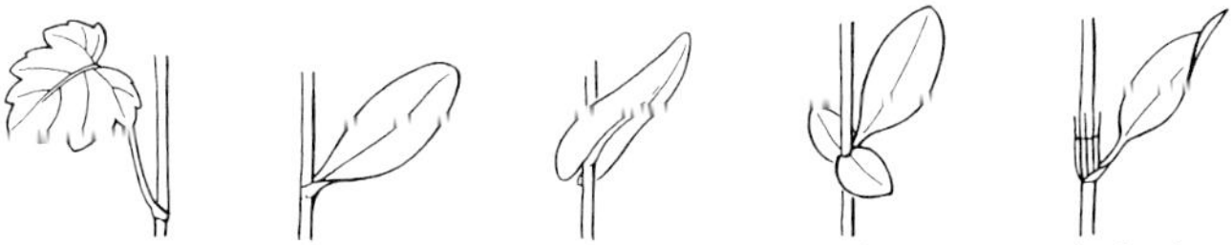

Venation

Leaf arrangement

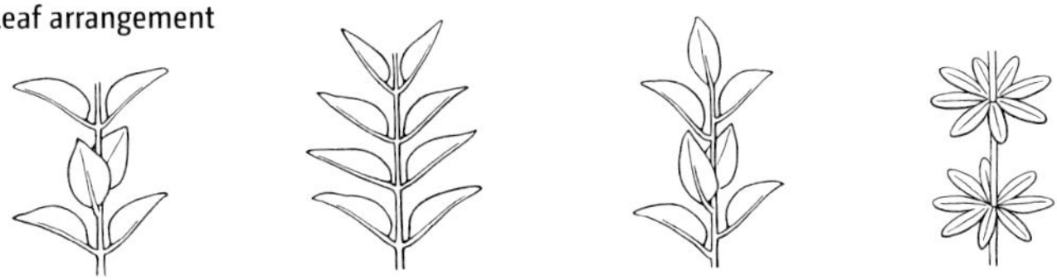

Underground plant parts

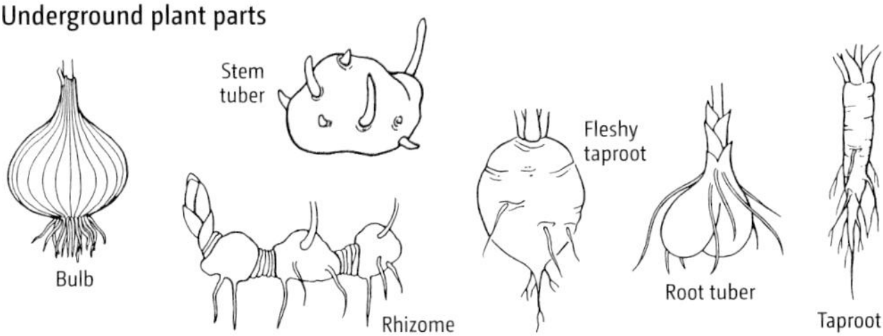

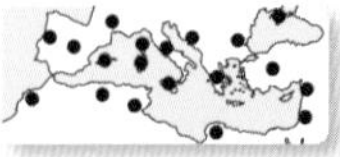

Fragrant Virgin's Bower

Clematis flammula · Ranunculaceae | May–Aug

The Mediterranean clematis species are woody at the base, with stems and leaves climbing to several metres in height. They have 4 petal-like sepals. Two rarer purple-flowered species of damp sites are particularly attractive. The fresh parts of clematis are poisonous.

Identification 3–5m. Leaves deciduous, opposite, mostly doubly pinnate, their leaflets long-stalked, entire or 2- to 3-lobed. Inflorescence paniculate, the scented flowers erect, about 2cm across, the 4 white sepals with felty hairs on the margins only.
Habitat Maquis, hedges.
Other species In contrast, the Early Virgin's-Bower has evergreen leaves. Flowers 1–4, 4–7cm across, bell-shaped, drooping, white to yellowish, sometimes with red marks, the outside generally hairy (Oct–Apr, Mediterranean).

Early Virgin's-Bower
Clematis cirrhosa

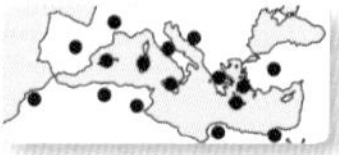

Spineless Caper

Capparis orientalis · Capparaceae | May–Oct

Everyone knows capers as something edible and by their characteristic taste. But few people know that capers are the unopened flower buds of a decorative plant preserved in salt or vinegar.

Identification 0.3–3m. Prostrate or hanging shrub. Leaves rather fleshy, hairless, rounded egg-shaped, blunt or with a notched tip, with two bristly short-lived stipules at the base. Flowers single on long stalks from the leaf axils, 5–8cm across. 4 sepals and 4 rather unequal petals, numerous purple filaments, the ovaries long-stalked, projecting.
Habitat Rocks, cliffs and walls, often cultivated.
Other species Capers, formerly called *Capparis spinosa*, are now divided into ten species, including the Jerusalem Caper: erect shrub with curved, decurrent thorny stipules (Mediterranean).

Jerusalem Caper
Capparis zoharyi

The fruits have a feathered hairy tail.

The buds open into impressive flowers.

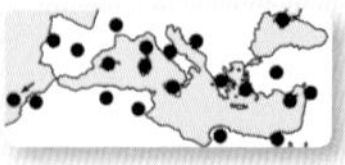

Sweet Alyssum

Lobularia maritima · Cruciferae | Jan–Dec

This species, which grows as a perennial in its native habitat, is also cultivated in a range of varieties further north in Europe as an annual bedding plant and has now become established in the wild.

Lobularia libyca

Identification 0.1–0.4m. A common species with much-branched stems becoming woody at the base. Leaves entire, unstalked, narrow lance-shaped, ± greenish grey from flattened forked hairs. White or pale pink honey-scented flowers in leafless racemes. The 4 petals, about 3mm in length. Siliculas 2–3.5mm, pointed egg-shaped with persistent style and 2 cells, usually with 1 seed in each cell.
Habitat Rocky and sandy coasts, roadsides, walls.
Other species *Lobularia libyca* is an annual. Leaves narrowed, stalk-like. Inflorescence leafy at base, the petals only 1–2mm long. Siliculas 3–5mm, with 2–6 seeds in each cell (Oct–May, southern Mediterranean, Canaries).

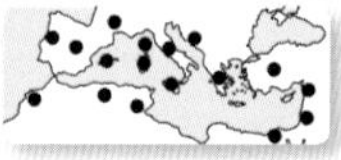

White Wall-rocket

Diplotaxis erucoides · Cruciferae | Jan–Dec

This is the only white-flowered species in the genus, all the others have yellow flowers.

Evergreen Candytuft
Iberis sempervirens

Identification 0.1–0.6m. Annual or biennial, branched, roughly hairy species, mostly with a basal rosette of lyre-shaped pinnately cleft leaves, those on the stem simpler, at base narrowed to almost spear-shaped. Flowers with 5–13mm long, white, purple-veined petals. Fruit ± spreading, linear, 2–5cm in length, beaked.
Habitat Cultivated land, often in vineyards, waste ground.
Other species Evergreen Candytuft forms small subshrubs with ± leathery leaves. Outer petals 7–11mm, much longer than the inner ones. Siliculas broad egg-shaped, 6–8mm, with wings widening towards the top, deeply notched (0.1–0.25m, May–July, rock crevices, scree slopes, also an ornamental plant, Mediterranean).

Each fruit cell contains only 1 hot-tasting seed.

The white petals often turn violet at the end of the flowering period.

Flax-leaved Daphne

Daphne gnidium · Thymelaeaceae | Jun–Oct

The species has the typical flower form of the Thymelaeaceae. A petal-like, 4-pointed calyx turns into a hypanthium of the same colour at its base. There are no petals.

Identification 0.5–2.5m. Stems erect, covered with uniform dense, bluish-green, 2–5cm long, tapering linear leaves, bare above with small glands on the underside, persisting for one year. Flowers 2–4mm in length, yellowish white, between the leaves at the ends of the shoots. Fruits spherical, 7–8mm, bright red.
Habitat Maquis and woods.
Other species The leaves of the Spurge Laurel are evergreen, pointed egg-shaped, up to 12cm in length and crowded at the ends of the stems. Flowers greenish-yellow, 7–12mm long. Fruits 8–10mm, pointed egg-shaped, black (Feb–May, broadleaved woods, S and W Europe, NW Africa).

Spurge Laurel
Daphne laureola

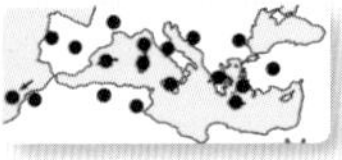

Tree Heath

Erica arborea · Ericaceae | Dec–Jul

The branches of this common species are still made into brooms while the beautifully veined, hard, heat-resistant wood of the roots is used for making pipe bowls (Bruyere Pipes).

Identification 1.5–4(–15)m. Evergreen shrub or small tree, young shoots covered in dense white hairs. Leaves needlelike, 3–5mm long, in whorls of usually 4, hairless, the underside completely covered by the inrolled leaf margin. Corolla white, 2.5–4mm, bell-shaped with 4(5) lobes. Base of anthers with spurs, stigma white.
Habitat Evergreen forest, maquis, mainly on acid soils. As far as the mountains of Central Africa.
Other species The branches of the Besom Heath are ± bare, the lower surface of the leaves only two-thirds covered by the inrolled edges. Corolla yellowish, anthers without spurs, stigma red (May–Jul, western Mediterranean).

Besom Heath
Erica scoparia

The berry-like stone fruits are extremely poisonous like all the plant parts.

A characteristic feature of the Tree Heath is the spreading, white-haired, young shoots.

Olive

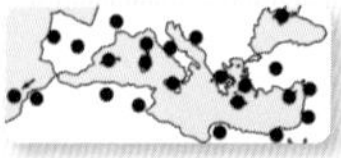

Olea europaea · Oleaceae | May–Jun

Olive oil, pressed from the ripe fruits, is an important economic factor for Mediterranean countries as cooking oil, for technical purposes and also in pharmaceutics. Table olives come from low-oil varieties and have to undergo a process of debittering before marketing, marinated in a saline solution or in oil.

Identification 2–15m. Evergreen tree, becoming very gnarled with age. Leaves opposite, short-stalked, oblong lance-shaped, leathery, dark green above, silvery grey below. 4-lobed flowers with short corolla tube in paniculate inflorescences. Fruits fleshy with a hard stone, green at first, brownish-red to bluish-black when ripe. **Habitat** A common cultivated tree throughout the Mediterranean region. The wild plants growing in the maquis with smaller leaves, thorny twigs and small low-oil fruits are classified as ssp. *oleaster*.

Olive harvest time November–March

Manna Ash, Flowering Ash

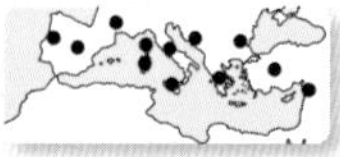

Fraxinus ornus · Oleaceae | Apr–Jun

Manna is the name given to the sap of this species of ash – obtained by making cuts in the bark – which is sweet, whitish and hard when dried. It is used as a gentle laxative amongst other things. The manna from the Bible is not identical, and there is no definitive conclusion about its source.

Identification 6–15m. Deciduous tree with opposite, pinnate leaves, the 5–9 leaflets distinctly stalked, pointed oval lance-shaped, irregularly serrate. Scented flowers in profuse panicles, with 4 linear 6(–10)mm long petals joined in pairs. Fruits tongue-shaped, pendulous, 2–4cm. **Habitat** Mixed deciduous woodland, up to the montane zone.

Other species The Narrow-leaved Ash has no corolla. Leaflets stalkless as in the Common or European Ash but with the same number of serrate teeth and lateral veins (Nov–Jan, damp deciduous forests, Mediterranean).

Narrow-leaved Ash *Fraxinus angustifolia*

The scented, whitish-yellow flowers are 4–7mm wide at the limb.

Only the Manna Ash has conspicuous white petals.

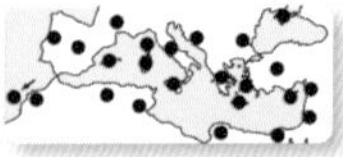

Sea Knotgrass

Polygonum maritimum · Polygonaceae | Mar–Dec

The Polygonaceae are characterised by a sheath around the stem at the leaf base, in this species reddish-brown below, silvery and transparent above with a deep slit and 8–12 obvious veins.

Identification 0.1–0.8m. Perennial beach plant, ± woody at base with spreading, prostrate to ascending stems. Leaves stalkless, greenish grey, egg- or lance-shaped, mostly inrolled at the edges. Flowers with 5-parted, whitish to pink, 3–4mm long, simple involucre, single or 2–4. Nuts triangular, shiny brown.
Habitat Sand and shingle beaches.
Other species The Horsetail Knotweed is easy to recognise with its shoots which resemble Horsetail (*Equisetum*). Leaves linear, soon dropping. Flowers 1–4 in loose spike-like inflorescences (roadsides, damp ruderal sites, southern Mediterranean, Canaries).

Horsetail Knotweed
P. equisetiforme

Prickly Drypis

Drypis spinosa · Caryophyllaceae | Jun–Sep

The little stems are tightly packed into a cushion. Symmetrical, dense branching results in strange, rectangular inflorescences whose flowers are surrounded by prickly bracts.

Identification 0.1–0.3m. Hairless perennial plant with much-branched, 4-angled, stiff, brittle stems. Leaves opposite, unstalked, lance-shaped to awl-shaped, rigid and spiny, channelled on the upper side, 2–3cm long. Flowers with 5 white or pink, long-clawed petals deeply divided into 2 narrow lobes. Outer, spiny bracts in the illustrated ssp. *jacquiniana* broadly lance-shaped, flowers scarcely projecting, in ssp. *spinosa* bracts narrow lance-shaped, distinctly longer than the flowers.
Habitat Ssp. *jacquiniana* on sandy and rocky coasts of the north-eastern Adriatic, ssp. *spinosa* in the mountains of Central Italy and the Balkan Peninsula.

Prickly Drypis with pink flowers

The shiny brown, 3-angled nutlets are roughly as long as the perianth.

Viewed from above the flower heads appear flat and rectangular.

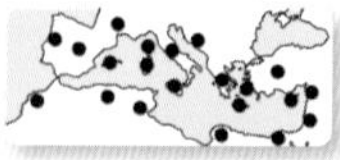

White Mignonette

Reseda alba · Resedaceae | Apr–Sep

The petals in Mignonette are usually split at the ends, so that at first glance it's difficult to see how many there are. There are often 5 or 6, the upper sometimes smaller and simpler.

Identification 0.3–0.9m. Annual to perennial plant, stem up to inflorescence with pectinate pinnatisect rough leaves, often with wavy edges. Flowers in dense racemes, 5(6)-parted. Petals white, about 6mm longer than the lance-shaped sepals, divided into 3 narrow points in their upper halves.
Habitat Roadsides, ruderal areas.
Other species Corn Mignonette or Rampion Mignonette has undivided leaves with up to 3 lobes. Petals 6, whitish, upper with numerous linear segments. Sepals finally extended to 13mm. Pendulous, blunt 3-angled, club-shaped capsules (0.1–0.5m, Mediterranean).

Corn Mignonette or Rampion Mignonette
Reseda phyteuma

Orange

Citrus sinensis · Rutaceae | Apr–Oct

The Orange, originating in SE Asia, has been cultivated in the Mediterranean region since the 16th century where it is an important economic factor along with other citrus species and their hybrids. Besides the fruits, they can also be distinguished by their leaf stalks which are ± broadly winged, depending on species.

Identification 2–5m. Evergreen tree, often with single, flexible thorns in the leaf axils of young twigs. Leaves dark green, broadly elliptical, pointed, indistinctly crenate. Stalk with narrow wings, inverted lance-shaped in outline. Flowers strongly scented, mostly with 5 rather thick, white petals and about 20 stamens.
Other species The leaves of the Lemon are crenate-serrate and clearly separated from the scarcely winged stalk. Flowers with a reddish tinge on the outside and 25–40 stamens (probably in the Mediterranean region since the 10th century).

Lemon
Citrus limon

The 4-angled, oblong, erect capsules are open at the top.

Orange fruits with their sweet flesh are spherical to egg-shaped.

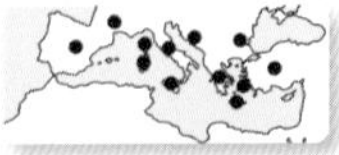

Almond-leaved Pear

Pyrus spinosa · Rosaceae | Mar–May

The Almond-leaved Pear was previously called *Pyrus amygdaliformis* on account of the similarity of its leaves to those of the Almond. It differs from the Common or European Pear, *Pyrus communis*, mainly due to the small, spherical, hard, sour fruits which are non-edible.

Identification 1–6m. Deciduous shrub or small tree, irregularly squarrosely branched and frequently thorny, young shoots grey, hairy. Leaves stalked, ovate-lanceolate, more than twice as long as broad, with smooth margins or slightly toothed towards the tip, with small warts, leaves soon falling. Flowers 2–3cm across, umbelliform, in groups of 8–12 on felty haired stems. Fruits with persistent calyx, spherical, 1.5–3cm, yellowish-brown when ripe.
Habitat Open woods, scrub, rocky ground, ascending to the montane zone.

Fruit of the Almond-leaved Pear

Evergreen Rose

Rosa sempervirens · Rosaceae | May–Jun

Evergreen plants are characteristic of the largely frost-free climate of the Mediterranean region. The deciduous roses of more northern parts of Europe are replaced here by an evergreen species.

Identification 3–6m. Climbing evergreen rose, only scattered, slightly curved thorns. Leaves pinnate, leaflets hairless, leathery and shiny, finely serrate. Flowers 1–3 (–12), corolla white, 2.5–5cm across, with downy column. Sepals mostly entire, egg-shaped with tapered tip, covered with stalked glands like the flower stems. Hips reddish, only about 1cm.
Habitat Maquis, open woods, hedges.
Other species The Firethorn is an evergreen thorny shrub. Leaves entire, finely crenate-serrate. Flowers 0.5–1cm across, fruits crimson (hedges, scrub, ornamental shrub, S Europe, SW Asia, naturalised elsewhere).

Firethorn
Pyracantha coccinea

Almond-leaved Pear forms attractive, freely-flowering, spreading trees.

The terminal leaflet of the 3–5-parted leaves is larger and tapers to a long point.

Montpelier Cistus, Narrow-leaved Cistus

Cistus monspeliensis · Cistaceae | Apr–Jun

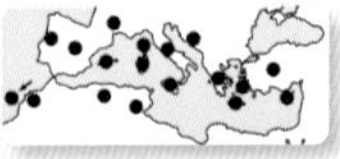

The species often forms large stands in garigue, recognisable even from a distance in the middle of summer due to the dark brown dried leaves. All species of Cistus have rather crumpled-looking petals.

Identification 0.3–1.2m. Aromatic, sticky glandular shrub. The opposite, stalkless leaves narrow lance-shaped, 4–8mm wide, with 3 main veins, dark green above, only sparsely hairy, with dense felty stellate hairs beneath and inrolled margin. Flowers 2–3cm across, yellow spotted at base, 2–9. Sepals 5, about equal, with ± dense long hairs.
Habitat Garigue, on acid soils.
Other species *Cistus clusii* has much narrower, 1–2mm wide, single-veined leaves. Sepals only 3, like the flower stems with long, white hairs (western Mediterranean to Sicily and S Italy).

Cistus clusii

Sage-leaved Cistus

Cistus salviifolius · Cistaceae | Mar–Jun

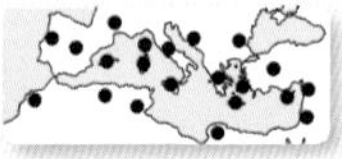

This species of Cistus has a wide distribution in the Mediterranean region. A few species in the western Mediterranean have even larger, 5–8cm wide, spectacular flowers sometimes with dark red markings at the base, such as the Gum Cistus.

Identification 0.2–0.9m. Aromatic but non-sticky, grey-green shrub. The opposite leaves short-stalked, ovate-elliptical, pinnately veined, strongly wrinkled and with stellate hairs on both sides. Flowers 3–5cm across, with yellow marking at base, 1–3. Five sepals.
Habitat Garigue, maquis, locally especially on acid rocks.
Other species The pleasant-smelling Poplar-leaved Cistus has long-stalked, pointed heart-shaped, generally hairless, poplar-like leaves. Flowers mostly larger, up to 6cm across, sepals also 5, all about equal (France, Iberian Peninsula, Morocco).

Poplar-leaved Cistus
Cistus populifolius

The hairless, spherical capsule opens from the tip with 5 teeth.

The 3 outer sepals are larger, pointed and heart-shaped at the base.

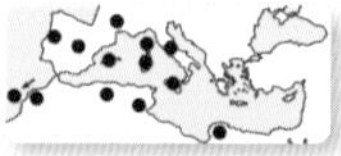

African Tamarisk

Tamarix africana · Tamaricaceae | Apr–Jun

Tamarisks often grow at the edge of water courses and on intermittently damp or even salty sites. The small flowers are 5- or 4-parted, depending on the species, with free petals.

Identification 1–3m. Shrub or small tree, leaves alternate, lying close to the twigs like scales, with ± transparent margins. Almost stemless, 5-parted flowers in 3–7cm long and 5–8mm wide catkin-like inflorescences on older twigs. Petals white or pale pink, 2–3mm long. Seeds with a sessile tuft of hair.
Habitat River courses, low coasts, also planted.
Other species The leaves of the Canary Island Tamarisk are covered with salt-secreting glands. Inflorescences narrower, only 1.5–5cm × 3–5mm, on current year's growth. Flowers pale pink outside, 5-parted (W Europe to Italy and Sicily, NW Africa, Canaries).

Canary Island Tamarisk
Tamarix canariensis

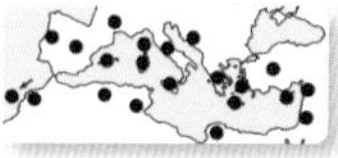

Myrtle

Myrtus communis · Myrtaceae | Jun–Aug

Myrtle plays an important part in Greek mythology. Even today, garlands or bouquets of myrtle are sometimes carried by the bride at a wedding as a symbol of love.

Identification 1–5m. Aromatic, evergreen, hairless shrub. The tough, short-stalked leaves opposite, at times in threes, pointed ovate-lanceolate and with transparent glandular dots. Flowers up to 3cm across, single in leaf axils, with rounded petals and numerous stamens.
Habitat Maquis, woods, an ornamental plant since antiquity.
Other species The frequently cultivated Fever Trees such as the River Red Gum also belong to the Myrtaceae. The fused sepals and petals in the bud opening via a little lid, releasing the numerous filaments (native of Australia).

River Red Gum
Eucalyptus camaldulensis

Scale-like leaves and flowers in catkins are typical of tamarisks.

The blackish-blue berries are topped by the long-lasting points of the calyx.

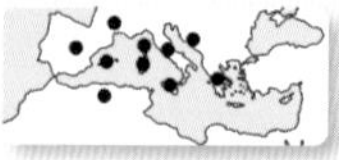

Prickly Samphire, Prickly Parsnip

Echinophora spinosa · Umbelliferae | Jun–Oct

The Prickly Samphire is a typical species of fixed dunes where it grows along with Marram Grass, Sea Holly and Sea Bindweed.

Seseli tortuosum

Identification 0.2–0.8m. Perennial, sturdy, grey-green, ± hairy, very thorny, fleshy beach plant. The stiff leaves are 2–3 pinnate, with keeled thorny segments beneath. Umbels of 5–12 hair-covered rays and 5–8-leaved, thorny involucres and involucels. Flowers with prickly sepals, the outer petals larger than the inner ones.
Habitat Sandy coasts.
Other species *Seseli tortuosum* has spreading branches with 3–4-fold pinnate, stiff, non-spiny leaves. Umbels with 4–11 rays, 0(–3) involucral bracts and 10 involucel bracts. Petals whitish to purple (dunes, pinewoods, roadsides, Mediterranean).

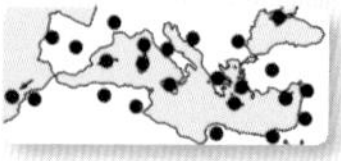

Shepherd's Needle, Venus' Comb

Scandix pecten-veneris · Umbelliferae | Apr–Jul

The Shepherd's Needle is still a common arable weed on calcareous soils in the Mediterranean region. It is threatened with extinction in more northern parts of Europe.

Southern Shepherd's-needle
Scandix australis

Identification 0.15–0.4m. Annual, hairy plant, the 2–4-fold pinnate leaves with narrow pointed lobes. Umbel of only 1–3 rays, no involucre, involucel bracts entire or with forward-pointing teeth. Outer flowers often larger. Fruit up to 8cm long, with strong, laterally highly compressed beak with bristly edges, distinctly separate from the seed-bearing part and much longer than this.
Habitat Cultivated land, waste ground.
Other species In the similar Southern Shepherd's-needle the involucel bracts often have a broad membranous margin. Fruits only up to 4cm long, beak not clearly separate from seed-bearing part (Mediterranean).

Even prior to flowering the plant is painful to stand on if walking bare-foot in the dunes.

The name Venus' Comb is derived from the comb-like fruits.

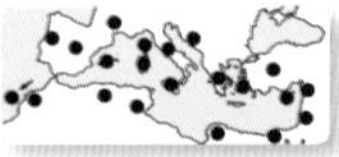

Toothpick plant

Ammi visnaga · Umbelliferae | May–Sep

The 30–50(–150!) rays of the large umbels thicken and harden towards fruiting time. They taste pleasantly spicy and in N Africa are dried and used as toothpicks.

Identification 0.2–1m. Easily recognisable, annual or biennial, large, hairless umbel. Leaves up to 3-times pinnate, with narrow linear to threadlike segments. Numerous pinnatisect involucral bracts and bristly involucel bracts. Hairless, ribbed, egg-shaped fruits 2–2.5mm long.

Habitat Damp ruderal sites, waste ground, cultivated and established in the wild in some places.

Other species In the Bullwort or Bishop's Weed, the leaf segments are broader with lance-shaped, cartilaginous toothed, blunt ends. Umbels of 15–30(–60) narrow rays, involucel bracts membranous (cultivated land, waste ground, roadsides, Mediterranean, Canaries).

Bullwort or Bishop's Weed
Ammi majus

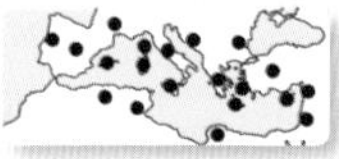

Mediterranean Hartwort

Tordylium apulum · Umbelliferae | Apr–Jun

Hartwort species have larger marginal flowers of a different shape and flat, rounded to elliptical fruits with a swollen edge. Of the 19 species in the Mediterranean region, most occur in the east.

Identification 0.2–0.5m. Annual plant with pinnate leaves, the lower with rounded-ovate, crenate-incised leaflets, the upper with entire, linear segments. Umbel rays 3–8, with small, awl-shaped involucral and involucel bracts. Characteristic, 4–8mm fruits with bladder-like hairs on the surface and whitish, crenate, finely papillose margin.

Habitat Cultivated land, waste ground, roadsides.

Other species The Egyptian Hartwort is distinguished by a purple floret in the centre of the umbel. Marginal flowers with 3 enlarged petals, the middle one 2-lobed (Cyprus, Turkey to Egypt).

Egyptian Hartwort
Tordylium aegyptiacum

At flowering time the umbels form domed hemispheres.

The marginal flowers have an enlarged, deeply 2-lobed petal.

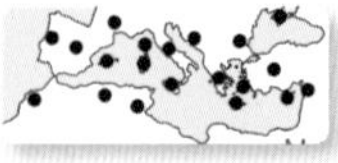

Strawberry Tree

Arbutus unedo · Ericaceae | Oct–Mar

The attractive, warty fruits of the Strawberry Tree turn from yellow to orange to a brilliant strawberry red. Contrary to expectations they have a bland taste (Latin *unedo*: I eat one), but are suitable for making jam, liqueur or spirits.

Identification 1.5–3(–12)m. Evergreen shrub or small tree with matt brown, finely fissured bark. Leaves tough and shiny, lance-shaped, serrate, 2–3 times as long as wide, green on both sides, with a stalk up to 1cm long. Flowers about 9mm, urn-shaped with recurved points.
Habitat Maquis, predominately on lime-poor soils.
Other species The Greek Strawberry Tree or Eastern Strawberry Tree flowers in spring. Trunk smooth reddish-brown. Leaves 1.5–3cm long-stalked, less than twice as long as wide, greenish-grey beneath, predominantly entire. Flower heads glandular hairy (Feb–Apr, eastern Mediterranean).

Greek Strawberry Tree, Eastern Strawberry Tree
Arbutus andrachne

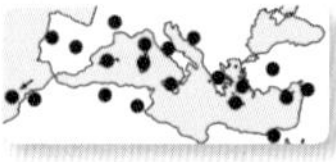

African Milkweed, Bristly-fruited Silkweed

Asclepias fruticosa · Apocynaceae | May–Sep

This poisonous plant containing cardiac glycoside is native to S and E Africa, but has long been known in the Mediterranean region as an ornamental, medicinal and agricultural plant (the silky seed hairs can be made into fine textiles).

Identification 1–2m. Only slightly woody, few-branched plant with milky sap. Leaves opposite or in whorls of 3, linear lance-shaped, ± inrolled at the edge. Flowers in overhanging stalked cymes, the approx. 10mm wide corolla with 5 recurved points and 5-parted fleshy corona. In the balloon-like capsule are numerous seeds with tufts of silky hair.
Habitat Naturalised in damp areas close to coasts.
Other species The Erect Milkweed has heart-shaped, pointed leaves. Flowers with spreading, blunt corolla lobes and small corona. Balloon capsules smooth (eastern Mediterranean).

Erect Milkweed
Cionura erecta

The Strawberry Tree is a characteristic feature of Mediterranean maquis.

The swollen fruits with beaked tips have soft spines.

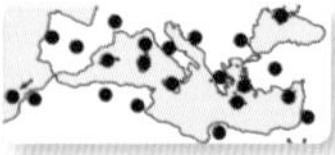

European Heliotrope, European Turnsole

Heliotropium europaeum · Boraginaceae | Feb–Nov

The Heliotrope gets its name from the ability to turn its leaves towards the sun. Some species are grown as ornamental plants on account of the vanilla-scented flowers.

Smooth Heliotrope or Seaside Heliotrope *H. curassavicum*

Identification 0.05–0.4m. Annual, green to grey plant with dense flattened hairs. Leaves stalked, ovate-elliptical, entire. Flowers stalkless, in bractless, simple or branched, free-flowering scorpioid cymes. Corolla with short tube and 2–4mm wide limb. Calyx divided to base, the readily spreading, externally shaggily hairy tips retained until the fruits are ripe.
Habitat Agricultural land and waste ground, roadsides.
Other species The Smooth Heliotrope or Seaside Heliotrope is prostrate, hairless and fleshy, with narrow spatula-shaped leaves. Flowers scented, 2mm wide at the limb (flowers Jul–Sep, Mediterranean, Canaries, native to America).

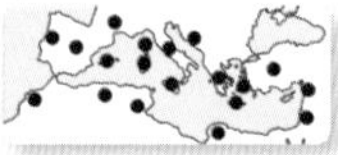

Laurustinus

Viburnum tinus · Viburnaceae | Jan–May

The typical shrub of maquis and evergreen woods is grown as an ornamental in a range of varieties.

Laurustinus fruits

Identification 1–3(–7)m. Evergreen, much branched shrub or small tree, young twigs reddish-brown. The opposite, stalked, leathery, entire leaves are pointed ovate-lanceolate, shiny dark green above, lighter and sparsely hairy below. Scented flowers in 4–9cm broad, dense corymbs at the ends of the twigs. Corolla 5–9mm across, with 5 rounded lobes and 5 white stamens. In autumn the 5–8mm egg-shaped, metallic blue to blackish-blue, slightly fleshy stone fruits are conspicuous, remaining a long time on the bush. Eating them is not recommended.
Habitat Shady, often damp woods.

The flowers of this species are scentless and have no appendages in the yellow throat.

The short fused petals are white inside and tinged pink outside.

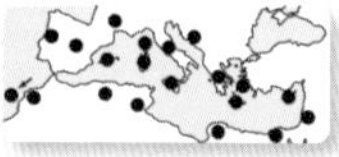

Ice Plant

Mesembryanthemum crystallinum · Aizoaceae | Mar–Jul

The spreading procumbent, fleshy plant occurs on salty sites near coasts. It was formerly used for the production of soda, but nowadays only occasionally as an addition to salads.

Identification 0.2–0.8m long. Annual plant, leaves spatula-shaped to broad egg-shaped, flat, ± red tinged, the lower stalked. Flowers ± stalkless, 2–3cm across, the numerous, very narrow white or pale pink petals joined at the base into a short tube, longer than the 5 sepals.
Habitat Naturalised species, native to S Africa.
Other species The Slender-Leaved Ice Plant has cylindrical leaves and only 1.5cm wide flowers. Petals white, yellow at the base, shorter than the sepals (Mediterranean, Canaries).

Slender-Leaved Ice Plant
M. nodiflorum

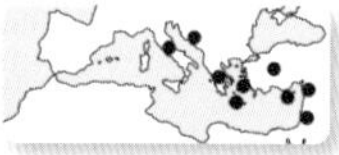

Storax

Styrax officinalis · Styracaceae | Mar–Jun

The pleasant-smelling resin from the bark (solid storax) is still used for producing smoke. On the other hand, the viscous balsam from the Sweetgum used primarily for perfumery is called liquid storax.

Identification 2–4(–7)m. Deciduous shrub or small tree. The alternate, broad egg-shaped, entire leaves are green, becoming hairless above and greenish-grey with dense stellate felty hairs beneath. Stalked flowers 3–6, about 2cm long, with short corolla tube and 5–7 overlapping lance-shaped lobes.
Habitat Open woods, scrub, river banks.
Other species The Oriental Sweetgum from the Hamamelidaceae family, with its lobed leaves and hanging, spherical flower heads is reminiscent of the Plane Tree (riparian forest, river banks, Rhodes, SW Turkey, Cyprus).

Oriental Sweetgum
Liquidambar orientalis

The plant is covered in large gleaming papillae for water storage.

The style remains on the spherical, white woolly, 1cm fruit for a long time.

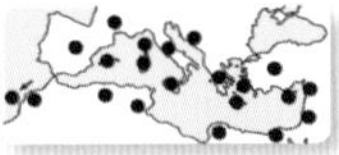

Branched Asphodel

Asphodelus ramosus · Asphodelaceae | Mar–Jun

The genus *Asphodelus* was botanically revised for the Mediterranean region, leading to significant changes in the nomenclature of the species.

Asphodelus cerasiferus

Identification 0.5–1.5m. Perennial plant with spindle-shaped thickened roots. Leaves in basal rosettes, 1–2(–4) cm wide, flat and slightly keeled. Inflorescence often much branched, pyramidal, on a strong stalk. Flowers 6-parted, sepals spread like a star, white with brownish-red central vein, 11–18mm long. Capsules ovate-elliptical, 5–13mm. Poisonous and not eaten by grazing animals, probably the commonest asphodel species.
Habitat Pastures, garigue, often forming stands.
Other species *Asphodelus cerasiferus* is similar but inflorescence less branched to simple, sepals 13–22mm long. Capsules much larger, spherical, 10–20mm (western Mediterranean).

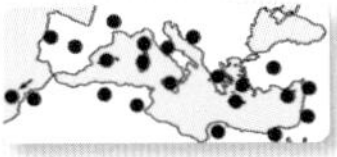

Hollow-stemmed (Onion-leaved) Asphodel

Asphodelus fistulosus · Asphodelaceae | Dec–Jun

The Hollow-stemmed Asphodel is the most common and robust of only 3 species with tubular leaves in a basal rosette and fibrous, non-thickened roots occurring in the Mediterranean.

Onion Weed
Asphodelus tenuifolius

Identification 0.2–0.7m. Annual to short-lived perennial, leaves hemicylindrical or cylindrical, somewhat rough at the edge and on the veins. The loose, simple or few-branched racemose inflorescence usually on a smooth stalk, with 6-parted flowers. The star-shaped spreading, 8–13mm sepals white to pale pink, with a reddish-brown central vein.
Habitat Roadsides, cultivated land, garigue.
Other species The cylindrical leaves of the Onion Weed are markedly rough and warty at the base on all the veins. Flowers bell-shaped, horizontally spreading, sepals 4–8mm long (southern Mediterranean, Canaries, SW Asia).

The asphodel meadows which were familiar to Homer are caused by overgrazing.

The basal rosette of leaves resembles chives but does not have an oniony smell.

Mountain Star-of-Bethlehem

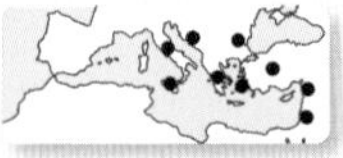

Ornithogalum montanum · Asparagaceae | Mar–May

Two groups can be distinguished amongst the many species of Star-of-Bethlehem in the region (all are poisonous bulbous plants): species with flowers in a cyme, the short stalk with only 2–15(–25) flowers, and species with 20–80 flowers in an elongated cylindrical raceme.

Identification 0.1–0.2m. The main identifying feature is the hairless, basal, 8–20mm wide, flat leaves with no central white stripe on the upper surface. Inflorescence a cyme of 5–15 flowers. Sepals 10–25mm long, with broad, green central stripe beneath.
Habitat Rocky ground, pastures.
Other species The Narbonne Star-of-Bethlehem belongs to the 2nd group. Flowers erect in a long raceme on ± equally long stalks. Leaves 4–6, persisting until after flowering (0.2–0.8m, fields, pastures, roadsides, Mediterranean, Canaries).

Narbonne Star-of-Bethlehem
O. narbonense

Sea Squill

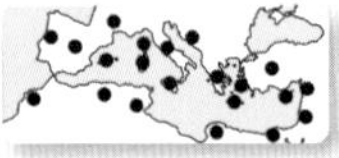

Drimia maritima s. l. · Asparagaceae | Aug–Oct

The poisonous bulb contains cardioactive glycosides which are fatal even in small doses. In the past the fleshy bulb scales were used for treating cardiac insufficiency. Nowadays specific extracts or the pure glycosides are only used in rare cases.

Identification 0.5–1.5m. The bulb which projects above the ground is white, up to 18cm across and up to 3kg in weight. The leaves, 30–100cm long and 3–10cm wide, appear in spring and are already dried up at the time of flowering. Flowering stalk with over 50-flowered, long and dense racemes. Flowers on ± erect stalks, their sepals 6–8mm long, white with green or purple central vein, spreading star-shaped, anthers greenish.
Habitat Pastures, rocky ground, garigue, sandy beaches.
Other species This variable species, formerly known as *Urginea*, is divided into several microspecies.

Bulbs with emerging leaves

Star-of-Bethlehem species take their name from the starry shape of the milk-white flowers

In autumn the tall flowering spikes tower above the parched surroundings.

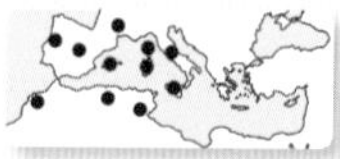

Neapolitan Garlic or Naples Onion
Allium neapolitanum

Three-cornered Garlic

Allium triquetrum · Amaryllidaceae | Feb–Jun

Leek species are perennial bulbous plants which can be identified by the smell of onion or garlic when crushed. The flowers are initially surrounded by a dry membranous involucre.

Identification 0.2–0.5m. Species easily identified by the sharply 3-angled flower stalk and the 2–3 hairless, keeled leaves with short above-ground sheath. Flowers 3–15 in a 1-sided cymose umbel, initially erect, later pendulous, the perianth segments white with green central vein, 10–18mm long, clustered, the tips curved outwards.
Habitat Damp, shady edges of thickets, ditches.
Other species In the Neapolitan Garlic or Naples Onion the flower stalk has 2 sharp edges and 1 rounded one. Cymose umbel almost spherical, formed from beaker-shaped to star-shaped spreading flowers (waste ground, maquis, Mediterranean, Canaries).

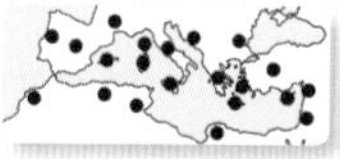

White Asparagus
Asparagus albus L.

Wild Asparagus

Asparagus acutifolius · Asparagaceae | Jul–Oct

In spring everywhere in the Mediterranean you come across local people on the hunt for young green asparagus shoots in woods and thickets. This wild asparagus is a popular dish.

Identification 0.4–2m. Climbing subshrub with squarrose, whitish twigs. In the axils of scale-like leaflets, bunches of 5–30 and more, roughly equal-sized, 2–8mm long, stiff, spiny short shoots (phylloclades). Flowers 1–4 with 3–4mm long, 6-parted, yellowish-white perianth. Berries black.
Habitat Woods, maquis, garigue.
Other species The White Asparagus has pale branches distinctly bent backwards and forwards. Leaves modified into stout, 5–12mm long thorns, with 10–20 soft, 5–25mm long, short-lived short shoots in their axils. Flowers 6–15, white, 2–3mm long, berries red (western Mediterranean).

The Three-cornered Garlic gets its name from the shape of its flower stalks.

The Wild Asparagus has no thorns but has spiny, short shoots.

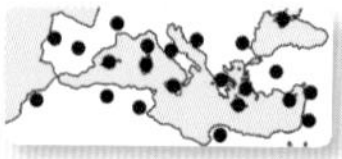

Butcher's Broom

Ruscus aculeatus · Asparagaceae | Oct, Feb–Apr

This plant has a variety of uses: the twigs are often used in dried flower arrangements, the young shoots are edible like asparagus. Extracts of the rootstock are used in medicines for venous diseases.

Identification 0.1–1m. Evergreen, branching, dioecious subshrub with leaf-like, spreading, up to 4cm long, broadly lance-shaped, stiff, spiny flat shoots arranged in 2 rows. Attached to these about 5mm wide, greenish-white to pale violet flowers with 6 sepals, often facing downwards due to rotation of the shoots. Shiny red poisonous berries.
Habitat In the undergrowth of maquis, evergreen and deciduous woods, up to the montane zone.
Other species The Spineless Butcher's Broom has unbranched stems with tongue-shaped, non spiny, 5–10cm long flat shoots (0.3–0.6m, Italy to Asia Minor).

Spineless Butcher's Broom, Mouse Thorn
Ruscus hypoglossum

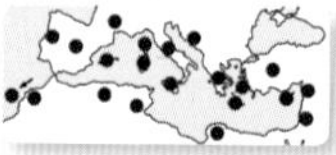

Rough Bindweed, Smilax

Smilax aspera · Smilacaceae | Aug–Dec

The widely distributed Rough Bindweed is very variable, especially in the case of the leaves which can be thorny or almost thornless. For this reason several subspecies and varieties have been described.

Identification Up to 6(–10)m high climbing, evergreen, hairless, dioecious shrub. Leaves leathery and shiny, narrow to broadly heart-shaped or spear-shaped, densely thorny or thornless at the edge and beneath on the main veins and likewise on the zig-zag-shaped stems. 2 tendrils at the base of the leafstalk. Inflorescences composed of small bunches of fragrant flowers with 6-parted, whitish or pink, 2–4mm long involucre.
Habitat Maquis and woods, also on walls.
Other species *Smilax aspera* ssp. *balearica*, with very narrow leaves, is considered an endemic subspecies or variety of the Balearic Islands.

S. aspera ssp. *balearica*

The small flowers can often be found by turning over the flat shoots.

The red berries can be found practically throughout the year.

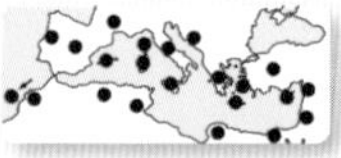

Bunch-flowered Daffodil, Polyanthus Narcissus

Narcissus tazetta · Amaryllidaceae | Dec–May

Daffodils flowering in gardens and parks are a harbinger of spring, but to see them in the wild is a particular pleasure. All species contain poisonous alkaloids.

Identification 0.2–0.6m. Bulbous plant with 3–6 blue-green, 0.5–1.5(–2.5)cm wide, linear, bluntly keeled leaves which are as long or longer than the sturdy, flattened 2-angled flower stalk. Scented, unequally long-stalked flowers, 2–15, with narrow tube and 6 spreading white, more rarely yellow, segments. These 8–22mm long, broad egg-shaped, usually touching or covering each other. The bowl-shaped corona 3–7mm long, yellow or orange. Variable species.
Habitat Damp meadows, pastures, frequently cultivated.
Other species The Late Narcissus flowers in autumn with 1(–3) delicate flowers. Corona only about 1mm long (Oct–Dec, garigue, waste ground, Mediterranean).

Late Narcissus
Narcissus obsoletus

Sea Daffodil

Pancratium maritimum · Amaryllidaceae | Jul–Oct

The Sea Daffodil, with its conspicuous, fragrant flowers in late summer in the dunes of the Mediterranean region, is threatened by seaside tourism.

Identification 0.2–0.6m. Beach plant with large bulb and 5–6 linear, blunt, twisted leaves. A cymose umbel of 3–10 flowers at the end of the compressed flower stalk. Flower with green, 5–8cm long tube and white, 3–5cm long, narrow perianth segments. The funnel-shaped corona about two-thirds as long, with 12 teeth. Fruiting stem bent to the ground.
Habitat Coastal dunes.
Other species The Illyrian Sea Daffodil has only a 1.5cm long corolla tube. Corona less than half as long as the perianth segments, deeply divided into 6 two-toothed segments (Apr–May, rocky ground, Corsica, Sardinia, naturalised on Elba and Capraia).

Illyrian Sea Daffodil
Pancratium illyricum

The Bunch-flowered Daffodil often forms large stands in meadows.

The seeds are jet-black and buoyant.

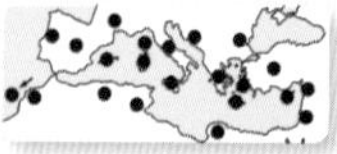

Annual Daisy

Bellis annua · Compositae | Feb–Jun

Daisies also occur in the Mediterranean region. Besides some rare endemics, the two following annual or perennial species are common and widely distributed. The Daisy or Common Daisy (*Bellis perennis*) familiar elsewhere in Europe also occurs.

Identification 0.02–0.15m. Delicate, annual, only sparsely hairy plant with fine roots. Stem leafy at base, without a clear rosette. Leaves spatula-shaped, crenate-serrate or entire. Flower heads on thin stalks, 0.5–2cm across, with 2 rows of involucral bracts, yellow disc florets and white ray florets.
Habitat Grassy, intermittently damp sites.
Other species The Southern Daisy or Wood Daisy is a perennial. Leaves with 3-veined lamina, all in a basal rosette. Heads long-stalked, 2–4cm across (0.1–0.3m, Sep–May, scrub, pastures, Mediterranean).

Southern Daisy, Wood Daisy
Bellis sylvestris

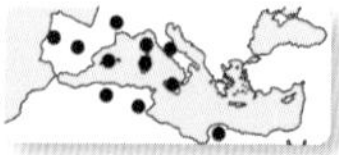

Seaside Chamomile

Anthemis maritima · Compositae | May–Sep

The roughly 100 species of the genus *Anthemis* – of which many only occur in the eastern Mediterranean – are not always easy to distinguish.

Identification 0.1–0.7m. Aromatic, woody at base, ± hairless, prostrate to ascending plant with strong stems. The green, fleshy leaves, with glandular dots below, 1–2 times pinnatisect, the topmost with only a few teeth or entire. Heads on scarcely thickened stalks, 1.5–4cm across, with yellow disc florets and white ray florets. Several rows of membranous-edged involucral bracts.
Habitat Sandy coasts.
Other species The Woolly Chamomile, in contrast, is covered in woolly white felty hairs. Heads on later thickened stalks. Variable species (May–Jun, sandy coasts, waste ground, more rarely on rocky sites, Italy to Turkey).

Woolly Chamomile
Anthemis tomentosa

You have to look carefully to find this species.

This species is easy to recognise from its fleshy, pinnatisect leaves.

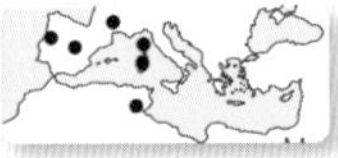

Goat's Thorn

Astragalus tragacantha · Leguminosae | Apr–Jun

The large genus *Astragalus* has its centre of distribution in SW to Central Asia. Besides thorny shrubs, the Mediterranean is also home to annual and perennial species.

Identification 0.1–0.4m. Low, spiny cushion-forming shrub. Leaves with 6–12 pairs of leaflets, these soon falling, leaflets oblong, covered below with dense hairs, each attached at its middle (hand lens needed!). Flower clusters of 3–8 pea flowers with 12–17mm long, usually white standard and pale violet keel.
Habitat Garigue close to the coast.
Other species The herbaceous Iberian Milk-vetch is up to 1m high and thornless. Flowers in a one-sided raceme with white, 25–31mm long corolla, calyx bell-shaped, reddish-brown or greenish (Feb–Apr, scrub, 2 subspecies, one in the western and one in the eastern Mediterranean).

Iberian Milk-vetch
Erophaca baetica

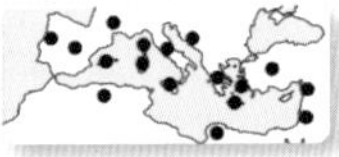

Canary Clover, Hairy Canary Clover

Dorycnium hirsutum · Leguminosae | Apr–Jul

The Canary Clover has a wide distribution in the Mediterranean.

Identification 0.2–1m. Shaggily white-haired subshrub or shrub. Leaves almost like fingers, 5-parted, stalkless or with very short rachis, leaflets obovate-oblong. Flowers 4–11 in short-stalked heads, 1–2cm long, standard as well as wing white to pink, keel with dark red, blunt tip. Calyx teeth even.
Habitat Garigue, maquis, open woods.
Other species The very variable, grey-green Badassi or Prostrate Canary Clover has flattened silky hairs, the 5 narrower leaflets stalkless, without rachis. Flowers 3–7mm long, white, in heads of 5–25, calyx weakly 2-lipped. Pods 3–5mm, ovate-spherical (Mediterranean, rarely to Central Europe).

Badassi or Prostrate Canary Clover
D. pentaphyllum

The rachis bears sturdy thorns.

The brown, 6–12mm long pods have a long-lasting style.

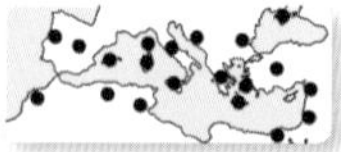

Cat-thyme Germander

Teucrium capitatum · Labiatae | Apr–Aug

Teucrium **species are easy to recognise: they lack the upper corolla lip, the lower lip has 5 lobes, usually with an extended middle lobe.**

Identification 0.1–0.4m. Small subshrub, overall densely felty with grey branched hairs. The ± stalkless, narrow oblong leaves often inrolled at the edge and with rounded teeth. Flowers in dense, less than 1cm wide heads, the corolla about 4mm, without a ring of hair in the throat. Calyx divided into 5 rather unequal teeth which are hidden beneath hairs. Several subspecies.
Habitat Stony grazing, garigue, open woods.
Other species The Ground-pine Germander is only slightly woody and usually glandular hairy. Leaves divided into 3–5 linear lobes. Flowers in a loose, one-sided raceme, corolla 10–15mm long, white, pink or light blue (western Mediterranean).

Ground-pine Germander
T. pseudochamaepitys

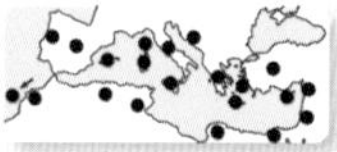

White Hedge-nettle, Great Hedge-nettle

Prasium majus · Labiatae | Feb–Jun

The Hedge-nettle is the only species in the region which produces fleshy, black fruitlets, differing in this from all other Labiatae which produce dry nutlets.

Identification 0.3–1m. Squarrosely branched, often climbing, hairless or sparsely hairy shrub. Leaves stalked, pointed egg-shaped with heart-shaped to truncated base, shiny dark green, toothed or crenate at the edge, the upper floral bracts sometimes entire. Flower whorls of only 1–2 short-stalked, white, often lilac-patterned, 17–23mm flowers. A ring of scale-like hairs in the corolla tube. Upper lip curved, entire, lower lip 3-parted with large middle lobe. Calyx 10-veined, at fruiting expanded to 25mm, weakly 2-lipped, with 5 broad, short-awned lobes.
Habitat Garigue, maquis, evergreen woods.

Fruits of the White Hedge-nettle

Cat-thyme Germander has white, cream-coloured, pink or purple flowers.

White Hedge-nettle grows mainly in coastal regions.

Winter Savory

Satureja montana · Labiatae | Jul–Sep

The two species of Savory with appetite-stimulating and digestive properties and their characteristic flavour are particularly good as a herb for dishes containing beans.

Summer Savory
Satureja hortensis

Identification 0.1–0.4(–0.7)m. Varied dwarf shrub. The leathery, lance-shaped leaves which are broadest in the middle, tapering to a sharp point, have dark glandular dots and short bristles at the edge but are hairless on the surface. Flowers 3–7 in small, stalked, one-sided false whorls, the 2-lipped corolla 6–14mm long, white, pink or violet, calyx throat hairy.
Habitat Rocky ground, grasslands.
Other species Summer Savory is an annual. Leaves soft, narrow linear and blunt, flowers 4–7mm long, whitish to violet (S Europe, Anatolia, also cultivated).

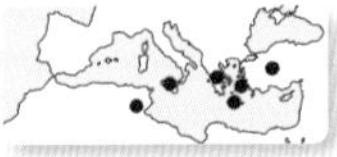

Cretan Oregano

Origanum onites · Labiatae | Apr–Aug

The Cretan Oregano is a medicinal plant and culinary herb and is sold as "oregano" along with other species. These include Greek Oregano, a Mediterranean sub-species of the very variable Wild Marjoram which also occurs elsewhere in Europe but which has little taste when dried and is less suitable as a culinary herb.

Greek Oregano
O. vulgare ssp. *hirtum*

Identification 0.3–1m. Aromatic subshrub with simple and glandular hairs as well as plentiful stalkless glands. Leaves heart or egg-shaped, occasionally finely serrate at the ends. Flowers with light green, overlapping, egg-shaped, about 3mm long bracts and white, 3–7mm long flowers, calyx 2–3mm, 1-lipped.
Habitat Rocks, garigue, banks.
Other species Greek Oregano has bristly hairy stems. Flowers with white or pink corolla in elongated inflorescences, calyx 5-toothed (eastern Mediterranean).

The Winter Savory has leathery, folded, sharply pointed leaves.

The flower heads of the Cretan Oregano are arranged in racemose umbels.

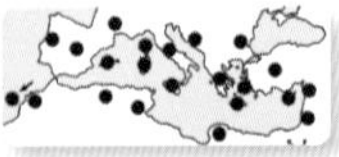

Trixago Bartsia, Mediterranean Lineseed

Bellardia trixago · Orobanchaceae | Apr–Jul

The plant occurs with white and, less commonly, yellow flowers. The latter easily leads to confusion with the yellow-flowered Yellow Bartsia (see p. 148). The calyxes of the two species are quite different.

Identification 0.1–0.8m. Annual, sticky glandular hemiparasite. Leaves opposite, stalkless, linear-lance-shaped, bluntly serrate at tips. Flowers in dense, 4-sided, spike-like inflorescence. Corolla 17–25mm long, with 2 ridges in the throat, lower lip white, upper lip pink or magenta or corolla completely yellow.
Habitat Garigue, pastures, waste ground.
Other species The Mediterranean or White Toadflax has alternate, linear leaves on erect, flowering shoots. Flowers in a loose raceme, corolla white to pale yellow, 12–22mm long, of which the narrow spur 8–13mm (0.2–0.5m, the Balearics, central and eastern Mediterranean).

Mediterranean or White Toadflax
Linaria chalepensis

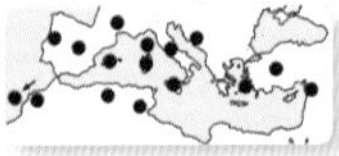

Bear's Breeches

Acanthus mollis · Acanthaceae | Apr–Aug

The leaves of *Acanthus* are said to have been the model for the ornamentation on the capitals of Corinthian pillars.

Identification 0.3–1.5m. Vigorous herbaceous plant with basal, long-stalked, pinnatisect and toothed leaves, the segments soft and thornless. Flowers in dense, cylindrical spikes, in the axils of finely thorny toothed bracts. The 3.5–5cm long corolla with a short tube and 3-lobed, white lower lip tinged with purple, the upper lip formed from the violet-tinged, enlarged section of the 4-part calyx.
Habitat Damp shady scrub, waste ground.
Other species The Spiny Bear's Breeches looks like a thistle with stiff 2(–3) times pinnatisect, thorny toothed leaves with white veins beneath (eastern Mediterranean).

Spiny Bear's Breeches
Acanthus spinosus

The calyx has 4 short teeth, is swollen and bell-shaped.

An enlarged calyx lobe projects above the corolla and replaces its upper lip.

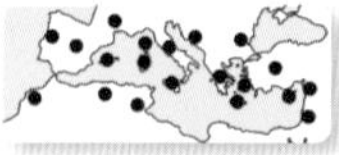

Osyris

Osyris alba · Santalaceae | Mar–Jul

As hemiparasites *Osyris* species obtain water and nutrients from the roots of other shrubs, but they can photosynthesise using their own chlorophyll.

African Sandalwood
O. lanceolata

Identification 0.4–1.2m. Low, leafless, dioecious shrub with switch-like branches. The evergreen, leathery, 1-veined leaves stalkless, linear lance-shaped, acute, 1–4mm wide. Flowers scented, with yellowish, simple 3(4)-parted perianth about 2mm long, the male flowers 3–6, the female flowers single. Bracts persistent during the flowering period.
Habitat Rocky ground, open maquis and woods.
Other species The African Sandalwood has 4–8(–14)mm wide leaves with weak striate venation and bracts which fall early. Fruits 7–10mm (1–4m, SW Mediterranean, Canaries).

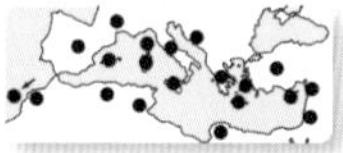

Cytinus Hypocistis

Cytinus hypocistis · Cytinaceae | Mar–Jun

There are 2 species of *Cytinus* in the Mediterranean region. They are obligate parasites on the roots of the Cistaceae.

Cytinus ruber

Identification 0.03–0.15m. Perennial plant without chlorophyll and with short, fleshy shoots which emerge from the soil and scale-like orange or red leaves. Flowers with bright yellow, simple, 4-lobed perianth. The species is parasitic on white-flowering *Cistus* and on *Halimium* and *Helianthemum* species; in the eastern Mediterranean also on the pink-flowering *Cistus parviflorus* and white-flowering *C. salviifolius*.
Habitat Maquis and garigue, always tied to the habitat of its host plants.
Other species *Cytinus ruber* is parasitic on pink-flowering *Cistus* species. Scale leaves crimson, perianth whitish (Mediterranean, Canaries).

The spherical, red 5–7mm diameter stone fruits are succulent and juicy.

The outer flowers are female, those in the centre male.

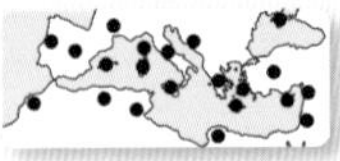

Bay Laurel

Laurus nobilis · Lauraceae | Feb–Apr

The aromatic, appetite-stimulating bay leaves are best known as a culinary herb. The species is the only native representative of the Lauraceae in the Mediterranean. The Lauraceae species *Persea americana* (avocado) from Central America is now cultivated as a fruit tree in the Mediterranean region.

Identification 2–10(–20)m. Evergreen tree or shrub. The alternate, dark green, leathery and hairless leaves stalked, oblong lance-shaped, tapering to a point at both ends, ± wavy at the edges. Flowers dioecious, with yellowish, 4-parted, 3–4.5mm long involucre in clusters of 4–6 in the leaf axils. The stone fruits up to 2cm across are bluish black when ripe (laurel berries). They are inedible.
Habitat Shaded damp woods, in the west only as an escapee from cultivation, planted as an ornamental and culinary herb.

Blackish-blue stone fruits of the Bay Laurel.

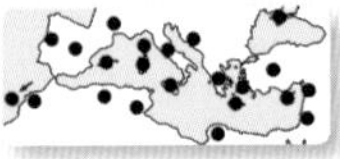

Hoary Mustard

Hirschfeldia incana · Cruciferae | Apr–Jun

A range of cruciferous plants with yellow flowers occur especially on cultivated land and waste ground, sometimes even forming complete stands. The species can be identified by their characteristic fruits.

Identification 0.2–1m. Annual, erect, branched plant. Leaves with grey hairs, lower lyre-shaped, pinnatisect, upper lance-shaped to linear. Flowers with pale yellow petals, the buds protruding. Siliquae lying against the stem, cylindrical, with the bulbous, easily shed beak 8–15mm long.
Habitat Agricultural land and waste ground, ruderal areas.
Other species The White Mustard is characterised by its spreading siliquae. These are bristly with a flat, sword-shaped beak. Several subspecies, cultivated for making mustard from the seeds (Mediterranean, Canaries).

White Mustard
Sinapis alba

The leaves smell pleasantly aromatic when crushed.

The siliquae lie close against the stem on thickened stalks.

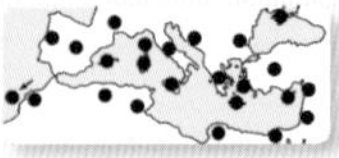

Yellow Horned-poppy

Glaucium flavum · Papaveraceae | Apr–Oct

While the fruits of poppy species are spherical to oblong capsules, Horned-poppy species are distinguished by often curved, horn-shaped, smooth or rough nodular siliquae.

Identification 0.2–0.9m. Biennial or perennial, blue-green frosted, almost hairless plant with milky sap. Lowest leaves stalked, pinnately parted, upper smaller, clasping the stem, stalkless, sinuately lobed. Flowers single, with 4 yellow, 3–4cm long petals, sometimes with a dark spot at base. Siliquae 15–30cm long, not narrowed towards the end.
Habitat Sandy and rocky coasts, ruderal sites, roadsides, carried inland.
Other species The annual Red Horned-poppy has bristly hairs. Petals scarlet or orange, often with a dark spot with light border at base (0.2–0.4m, Apr–Jul, Mediterranean, Canaries).

Red Horned-Poppy
Glaicium comiculatum

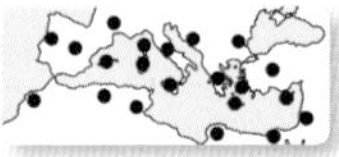

Hypecoum procumbens

Hypecoum procumbens · Papaveraceae | Feb–Jul

In the Papaveraceae, species of the genus *Hypecoum* are distinguished by their 3-lobed petals. Only the two sets of opposite petals are the same shape, so the flower has 2 planes of symmetry.

Identification 0.05–0.4m. Annual, hairless, green or grey-green plant with ascending or spreading procumbent stems with watery sap. Leaves 2–3 times pinnate, with ovate-lanceolate, pointed end segments. Flowers with 2 narrow sepals and 4 light- to golden-yellow petals, greenish on the underside. The external ones 7–10mm, about as long as wide, only weakly 3-lobed, the inner ones deeply divided into 3 narrow leaflets, the central one bent inwards. Siliquae often bent and segmented, 4–6cm long. Variable species.
Habitat Cultivated land, waste ground, sandy beaches, near the coast.

Fruits of *Hypecoum procumbens*

The bud is surrounded by 2 softly bristly sepals which are soon shed.

Hypecoum procumbens spreads over sandy ground.

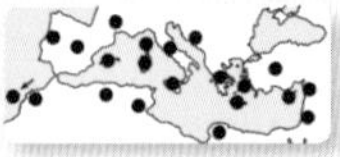

Fringed Rue

Ruta chalepensis · Rutaceae | Mar–Jul

Rues have a typical, strongly aromatic smell. Due to photosensitising substances, skin contact with fresh plant parts in the presence of sunlight can lead to inflammation.

Identification 0.2–0.8m. Hairless, herbaceous plant, woody at base. Leaves doubly pinnate, with oblong, blunt end segments. Inflorescence without glands, loosely cymose, flowers 1–2.5cm across. Petals yellow, distinctly fringed at the edges, sepals rounded. Capsules with dense sunken glands, about 5mm across, 4(5)-parted.
Habitat Garigue, rocky ground, roadsides.
Other species The similar Common Rue is known for its culinary and medicinal uses, but can have adverse effects. Petals wavy and with few irregular denticles, but not fringed (originally only on the Apennine and Balkan Peninsulas, but naturalised from cultivation).

Common Rue
Ruta graveolens

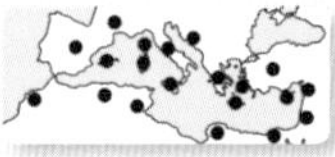

Spur flax

Thymelaea hirsuta · Thymelaeaceae | Oct–May

The loose shrub often with overhanging branches is unmistakeable due to its scale-like, rather fleshy leaves arranged like roof tiles.

Identification 0.4–1.5(–2)m. Leaves shiny dark green and hairless above, somewhat curved, white felty beneath like the young twigs. Flowers in heads of 3–6(–10) at the end of this year's shoots. Perianth simple, the petal-like, yellow, blunt 4-lobed calyx passing into a hypanthium.
Habitat Garigue, reaching as far as semi-desert.
Other species The leaves of *Thymelaea tartonraira* spread from the upper section of the twigs and are ± silkily hairy, in the illustrated ssp. *valentina* from Spain, narrowly oblong. Flowers 7–8mm long, surrounded by bracts (Mar–May, individual subspecies occur locally almost throughout the Mediterranean).

T. tartonraira

The central 5-parted flower is surrounded by 4-parted flowers.

The petal-like, blunt 4-lobed calyx is yellow and hairless inside.

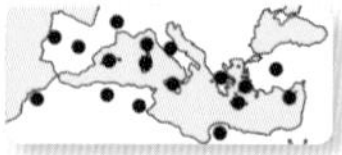

Autumn Buttercup

Ranunculus bullatus · Ranunculaceae | Aug–Feb

The species name *bullatus* refers to the form of the leaves which, broadly egg-shaped, coarsely crenate and hairy beneath, tend to swell upwards like blisters (Latin *bulla*: bubble).

Identification 0.05–0.3m. Perennial autumn and winter flowering plant. All leaves basal, spreading horizontally. Flowers with 7–18mm long, yellow petals and 5 greenish, hairy sepals. The 1–2mm long fruits in heads, smooth and hairless, with a tiny, somewhat curved beak.
Habitat Rocky ground, garigue, tree plantations.
Other species The Spiny-fruited or Rough-fruited Buttercup is a spring-flowering annual. Lower leaves long-stalked, rounded kidney-shaped, coarsely crenate, 3–5 lobed. Petals 4–11mm long. Fruits 7–8mm, with hooked beak, warty spines on the surfaces (damp sites, cultivated land, Mediterranean, Canaries).

Spiny-fruited or Rough-fruited Buttercup
R. muricatus

Yellow Flax

Linum campanulatum · Linaceae | Apr–Jul

Many flax species have surprisingly large flowers. Besides several yellow species, white, blue and red-flowered species are also eye-catching (see p. 214).

Identification 0.05–0.35m. Perennial, hairless flax species, woody at base, often with non-flowering rosettes. Stem with very narrow wings. Lowest leaves spatula-shaped, others lance-shaped, single-veined, with 2 glands at base. Flowers 3–5, short-stalked, petals 2–3.5cm long. Sepals expanded at fruiting, longer than the capsule, this with an approx. 2mm long beak.
Habitat Rocks, in gravel beds of dried-up streams.
Other species The Upright Flax has only 6–12mm long petals. Leaves linear-lanceolate, finely serrate at edge and therefore rough in contrast to other small-flowered annual species (grassy areas, garigue, maquis, Mediterranean, Canaries).

Upright Flax
Linum strictum

The flowers of this species have five or more petals.

The petals taper gradually towards the base.

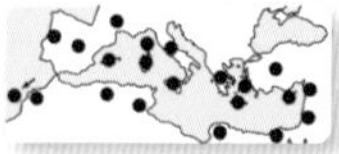

Bermuda-buttercup, African Wood-sorrel

Oxalis pes-caprae · Oxalidaceae | Dec–May

This species, native to S Africa (not Bermuda), has only occurred in the Mediterranean region since the start of the 19th century, having established in the wild from garden escapes. Its successful progress has been achieved by the proliferation of underground bulbils alone.

Identification 0.1–0.5m. Plant with a rosette of clover-like leaves, the leaflets deeply reverse-heart-shaped and hairy beneath. Inflorescence long-stalked, with 6–12 flowers arranged in an umbel, petals lemon yellow, 2–2.5cm long. Of the 3 possible flower forms with different lengths of styles and stamens, only one has reached the Mediterranean. This means that pollination is not possible and neither is the development of fruit capsules.
Habitat Cultivated land, especially below stands of trees, at flowering time dominating the landscape in some areas. Double flowers also occur.

Double cultivar of the Bermuda-buttercup

Perforate St John's-wort

Hypericum perfoliatum · Hypericaceae | Mar–Jun

Important features for identifying St John's Wort species are transparent, black or red glands and their distribution on the flowers and leaves. The filaments are usually in 3 or 5 bundles.

Identification 0.2–0.8m. Perennial, hairless, upright herb, leaves opposite, with transparent and black glands. Petals 8–12mm long, with black glands especially at the edges, sepals dentate-ciliate, with black glands and lines on the surface.
Habitat Damp shady ruderal areas, waste ground.
Other species The Shrubby St John's Wort is a small shrub with up to 3 whorled, linear, transparent glandular leaves inrolled at the edges. Petals 5–10mm long, sepals with black glands at the edges (Rocky areas, maquis, eastern Mediterranean).

Shrubby St John's-wort
H. empetrifolium

Flowers nodding in bud, later upright and funnel-shaped.

The ovate-lanceolate leaves partially clasp the stem at the base.

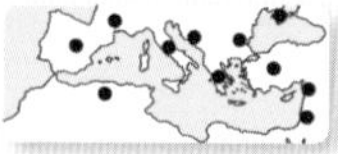

Jerusalem Thorn, Christ's Thorn

Paliurus spina-christi · Rhamnaceae | May–Sep

The twigs of this species were probably not used to weave Christ's crown of thorns. It is likely that the plant used for this was the Thorny Burnet (see p. 276) which is common in the region, or the tree-forming Christ's Thorn Jujube (*Ziziphus spina-christi*) is also a possibility.

Lotus Tree
Ziziphus lotus

Identification 2–3m. Deciduous shrub with zigzag-shaped, often hanging branches. Leaves arranged almost distichously, with unequal egg-shaped, indistinctly crenate-serrate, 3-veined lamina. Of the 2 thorny stipules one longer, straight, the other shorter and curved. Flowers golden yellow, 5-parted, about 2mm across, in small racemes borne in the leaf axils. Fruits flat, 1.5–3cm.
Habitat Deciduous forests, scrub, hedges.
Other species The Lotus Tree has similar twigs, leaves and flowers, but fleshy, roughly 1cm fruits (dry, stony sites, southern Mediterranean).

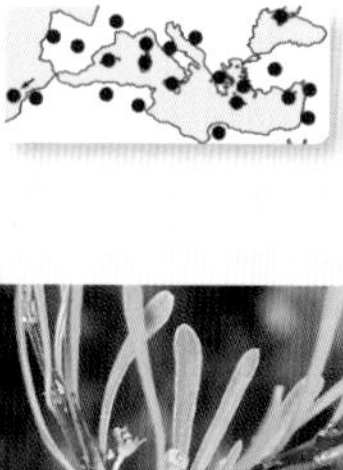

Italian Buckthorn

Rhamnus alaternus · Rhamnaceae | Mar–Apr

The alternate leaves of this species make it easily distinguishable from the similar Mock Privet (see p. 282 which has opposite leaves), even without the flowers or fruits.

Black Hawthorn
Rhamnus lycioides

Identification 1–3(–5)m. Evergreen, thornless, dioecious shrub. Very varied, leathery leaves, lance-shaped to egg-shaped, pointed or rounded, ± distally toothed or entire. Inflorescence racemose, dense. Flowers without corolla, the 4–6mm calyx greenish-yellow, with 5(4) lance-shaped, pointed lobes. Spherical, 4–6mm, slightly fleshy stone fruits.
Habitat Maquis, woods, also an ornamental shrub.
Other species The Black Hawthorn is deciduous or evergreen. Leaves likewise variable (ssp. *lycioides* illustrated), linear or reverse egg-shaped, mostly in bunches. Calyx with 4(5) lobes (Mediterranean).

The dried fruits have a broad wavy edge.

The red, then black fruits remain on the bush for a long time.

Yellow Rock-rose

Halimium calycinum · Cistaceae | Jan–Jul

Like the white, pink or crimson *Cistus* species, the predominantly yellow-flowered *Halimium* species are also called Rock-roses. Their capsules always have 3 valves, whereas *Cistus* has 5, 6 or up to 12.

Identification 0.2–0.6m. Subshrub, on non-flowering shoots with up to 3mm wide leaves resembling Rosemary, shiny green above, white felty beneath, with inrolled margins. Stipules zero. Flowers 2–3cm across, the 5 petals pale yellow, always without spots.
Habitat Fixed dunes, maquis, pinewoods.
Other species *Halimium halimifolium* is a much-branched shrub with silvery grey-green, oblong elliptical, 0.5–2cm wide leaves. Petals often with a dark spot at base. 5 sepals, of which the two outer ones very small (0.5–1.5m, Mar–Aug, western Mediterranean).

Halimium halimifolium

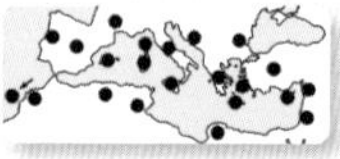

Spotted Rock-rose

Tuberaria guttata · Cistaceae | Mar–Aug

***Tuberaria* species are annual, more rarely perennial herbs. They form basal leaf rosettes which can be dried up by flowering time. The capsules have 3 valves.**

Identification 0.1–0.4m. Delicate, upright, hairy, variable species. Basal rosette leaves and lower stem leaves oblong elliptical, 1–3-veined, upper narrower, often with stipules. Inflorescence loosely hairy, the 1–2cm wide flowers thin and long-stalked, petals usually with dark brown spot at base. 3 large inner and 2 much smaller outer sepals.
Habitat Grassy areas, on sand, garigue.
Other species The woody *Tuberaria lignosa*, with a branched rhizome, is perennial. Leaves of the persistent rosette with 3(5) veins, white hairy beneath. Flowers 2–3cm across, without spots (western Mediterranean and Canaries).

Tuberaria lignosa

The Yellow Rock-rose has 3 hairless, sometimes red-tinged, pointed sepals.

The flower stems of the buds are bent backwards.

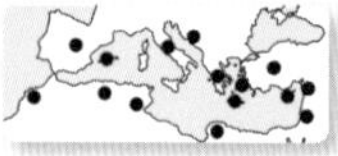

Lavender-leaved Rock-rose

Helianthemum syriacum · Cistaceae | Mar–Jun

Of the roughly 60 yellow, white or pink flowering low shrubs or subshrubs, more rarely annual species, of the genus *Helianthemum*, most are limited in their distribution to smaller sub-areas in the western Mediterranean.

Identification 0.1–0.5m. Small shrub with oblong "lavender-like" leaves, inrolled at the edge, grey felty above, white felty beneath with large stipules. Flowers 10–20mm across, in 3–5 times branched inflorescences. 3 inner sepals pointed egg-shaped, 2 outer, smaller, often ciliate. Capsules with 3 valves.
Habitat Rocky ground, garigue, open woods, on limestone.
Other species The name Cat's-head Rock-rose refers to the long spreading hairy calyx, which looks like a cat's head when in bud (coastal rocks, Spain, Balearics, Sardinia, NW Africa).

Cat's-head Rock-rose
H. caput-felis

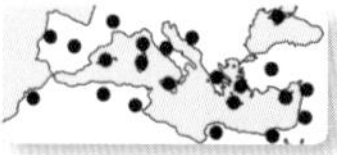

Thyme-leaved Rock-rose

Fumana thymifolia · Cistaceae | Apr–Jun

***Fumana* species are mostly dwarf shrubs with yellow flowers in which, in contrast to similar Cistaceae, the outer filaments do not have any anthers.**

Identification 0.1–0.3m. Sticky-glandular dwarf shrub. Leaves with stipules, lower narrowly oval, shorter than the needle-like upper leaves which are inrolled at the edges and opposite, at least to the middle, and alternate above this. Flowers 10–15mm across. The 3 inner sepals larger than the 2 external ones. Capsules with 3 valves.
Habitat Garigue, rocky ground, predominately on limestone.
Other species The Arabian Rock-rose has alternate, up to 5mm wide, ± glandular hairy, roughly equal-sized, oblong-elliptical leaves with stipules. They are distributed evenly along the stem. Flowers 12–20mm across (Mediterranean, absent from SW Europe).

Arabian Rock-rose
Fumana arabica

The young flower stems are rolled up like a snail and unfurl later.

The inner sepals have 4–5 prominent veins.

Squirting Cucumber

Ecballium elaterium · Cucurbitaceae | Mar–Sep

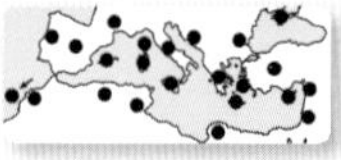

The fruits resembling cucumbers have an interesting dispersal mechanism: when ripe they detach themselves from their stalks through even gentle contact, shooting out their contents, a powerful skin irritant, along with the seeds. Watch out!

Colocynth or Bitter Apple
Citrullus colocynthis

Identification 0.2–1m long. Perennial, prostrate, rather fleshy, stiffly hairy plant without tendrils. Leaves stalked, heart-shaped to triangular, usually toothed and wavy. Flowers yellowish, 2–4cm across, deeply 5-parted, male in racemes, female single on stalks in the leaf axils. 3–7cm long, roughly hairy fruits.
Habitat Roadsides, waste disposal sites, waste ground.
Other species The Colocynth or Bitter Apple is also part of the cucumber family. It has conspicuous, highly poisonous, spherical, up to 12cm, green marbled or yellow fruits (southern Mediterranean, Canaries).

Rock Samphire

Crithmum maritimum · Umbelliferae | Jul–Oct

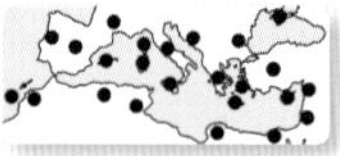

The structure of the fruits of the Rock Samphire help to spread it along the coasts. The ovary wall consists largely of a spongy, air-filled tissue which makes the fruit buoyant.

Rock Samphire leaves

Identification 0.1–0.6m. Hairless umbellifer, woody at base. The blue-green, fleshy leaves 2–3 times pinnate, with linear, pointed segments. Umbels of 10–32 sturdy rays, flowers with yellow-green scarcely 1mm long petals, inrolled at the tips. Involucral and involucel bracts 5–10, lance-shaped, membranous, recurved at fruiting time. Fruits 5–6mm, ovate-oblong, yellowish to reddish. The salty and rather bitter-tasting young leaves are still occasionally used in salads or prepared as a vegetable.
Habitat Mediterranean, Atlantic and Black Sea coasts.

After dropping its fruits, the hook-shaped stems remain for a long time.

Rock Samphire grows on rocky coasts, sometimes on walls in the splash zone.

Giant Fennel

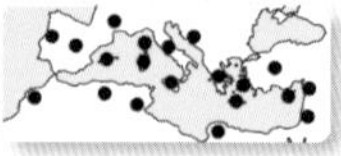

Ferula communis · Umbelliferae | Apr–Jun

Prometheus is said to have used a stem of the Giant Fennel to take the fire from Mount Olympus. The spongy pith inside is flammable and keeps smouldering without the stem burning away.

Identification 1–3m. Perennial herbaceous plant with many-pinnate, hairless leaves, the lower long-stalked, conspicuously large, the upper with inflated, leathery sheaths, leaflets linear and flat. Inflorescence many branched, the double umbels spherical, composed of 30–40 rays without involucre. Fruits compressed, with lateral wings.
Habitat Garigue, grazed land, often close to habitation.
Other species *Ferulago nodosa* is unmistakeable with its swollen nodes on the stem. Involucre and involucel of ovate-oblong leaflets. Fruits with rather wavy, lateral and narrower dorsal wings (to 1.5m in height, waste ground, rocky areas, SE Europe).

Ferulago nodosa

Shrubby Hare's-ear

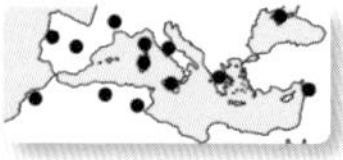

Bupleurum fruticosum · Umbelliferae | Apr–Sep

The Hare's-ear is easy to distinguish from other umbels with its simple, entire, often parallel-veined leaves.

Identification 0.6–2.5m. Aromatic, hairless shrub, often with reddish branches. Leaves almost stalkless, evergreen, leathery, blue-green beneath, the midrib ending in a small prickly point. 5–25 umbellets on vigorous rays with 5–6 recurved, short-lived bracts and bractlets. Ripe fruits 7–8mm, with narrowly winged ribs.
Habitat Rocks, garigue, also an ornamental shrub.
Other species The delicate Small Hare's-ear is an annual. Leaves lance-shaped, tapering to a point, with 3–5 parallel veins. Umbellets 2–5, unevenly long-stalked, bracts and bractlets upright, ± awned, almost hiding the flowers (0.05–0.3m, cultivated land, garigue, SW Europe).

Small Hare's-ear
Bupleurum baldense

This tall herbaceous plant often makes an impressive sight in the landscape.

Shrubby umbellifers like this Hare's-ear are a surprising sight in the Mediterranean.

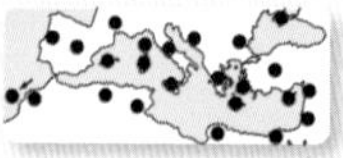

Wild Fennel

Foeniculum vulgare · Umbelliferae | Jul–Sep

Wild Fennel is often seen alongside streets and in ditches. In spring before flowering it can be recognised by the smell if you crush part of a leaf.

Vegetable fennel var. *azoricum*

Identification 0.5–2.5m. Biennial or perennial, tall, hairless, blue-green frosted umbellifer. Leaves 3–4 times pinnate, with fine, somewhat fleshy tips. The umbels of 4–45 unequally long-stalked umbellets without involucre and involucel. Sepals zero. Fruitlets each with 5 distinct but wingless ribs, with a rather strong, bitter taste in the case of the Wild Fennel (designated as ssp. *piperitum*). Various varieties are cultivated for harvesting the fruits: the milder, Sweet Fennel as a spice, the Bitter Fennel for medicinal purposes. The vegetable fennel forms thickened sheaths on the basal leaves like an onion.
Habitat Roadsides, waste ground, river courses.

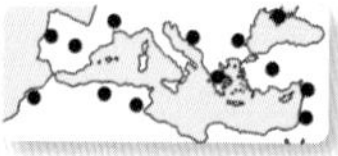

Wild Jasmine

Jasminum fruticans · Oleaceae | Apr–Jun

We expect Jasmine to have scented flowers, but this is not the case with the two yellow-flowered species shown here. For extracting perfume white-flowered climbers such as *J. grandiflorum* are cultivated.

Winter Jasmine *Jasminum nudiflorum*

Identification 0.5–3m. Shrub with slender, 4-angled, green branches. Leaves alternate, with 3 blunt leaflets, less frequently simple. Flowers single or up to 5 on short side branches. Corolla yellow, 8–20mm across, with long tube and (4–)5(–6) flat spreading, blunt lobes. Berries 7–9mm, turning black.
Habitat Garigue, open woods, also an ornamental shrub.
Other species The Winter Jasmine produces its pairs of opposite flowers on overhanging branches from December onwards before the leaves open. Corolla usually 6-lobed, 15–30mm across (ornamental shrub, established in the wild in France, native to China).

The flowers have blunt petals inrolled at the ends.

Opposite leaves are an exception in the Oleaceae.

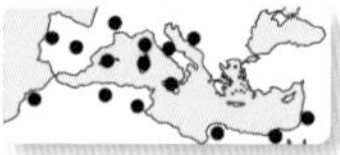

Sea Crosswort

Crucianella maritima · Rubiaceae | May–Sep

The Sea Crosswort has leaves arranged in whorls of four, the flowers are 5-parted.

Identification 0.1–0.4m. Prostrate-ascending plant with whitish, smooth stems, woody at base. The leathery, blue-green, white-edged leaves flattened or arranged like roof tiles, ovate-lanceolate with a spiny tip. Inflorescence spike-like, 1–5cm long, the 10–13mm long, narrow funnel-shaped, yellow flowers with 5 short, convergent tips, the free bracts distinctly projecting.
Habitat Fixed dunes.
Other species The annual Broad-leaved Crosswort has 4–6 spreading leaves. Inflorescence linear, about 15cm long, the 4-lobed 4–7mm long flowers somewhat longer than the basally-joined bracts (Apr–Jul, open woods, garigue, pasture, Mediterranean).

Broad-leaved Crosswort
Crucianella latifolia

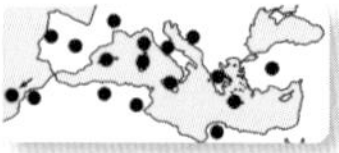

Wild Madder

Rubia peregrina · Rubiaceae | Apr–Aug

Mediterranean maquis can be almost impenetrable. Plants which contribute to this are those with prickles like the Wild Madder, Rough Bindweed, Evergreen Rose and species of Blackberry which climb towards the light.

Identification 0.5–2.5(–7)m. Climbing plant, woody at base. Stem overwintering, 4-angled, narrowly winged, often rough like the leaf edges and mid veins from small, backward pointing teeth. Leaves leathery, from narrowly to broadly ovate-lanceolate, 1-veined, in whorls of 4–8. The many-flowered inflorescences distinctly longer than the bracts. Corolla 4–8mm across, yellowish, usually with 5 awned lobes, calyx absent.
Habitat Maquis, woods, hedges, walls.
Other species The Narrow-leaved Madder is similar, but inflorescences scarcely longer than the bracts. Corolla lobes only tapering (eastern Mediterranean).

Narrow-leaved Madder
Rubia tenuifolia

The Sea Crosswort belongs to the summer-flowering species in the dunes.

The 4–6, spherical, succulent fruits later turn black.

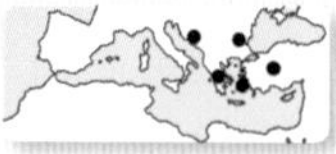

Onosma heterophylla

Onosma heterophylla · Boraginaceae | Apr–Jul

There are numerous *Onosma* species in the eastern Mediterranean, but many have restricted distributions. Some 20 species are known from Greece alone, but only 3 from Spain.

Identification 0.15–0.5m. Plant, woody at base. Basal and lower stem leaves oblong, narrower than the upper, with "star bristles" on small protuberances on both sides: several short bristles surrounding 1 longer one. Flowers ± stalkless, corolla pale yellow, with downy hairs, gradually narrowed towards the base, 20–30mm long.
Habitat Rocky ground, grasslands.
Other species The biennial Greek Golden Drop has conspicuous, 5–8mm long, white, simple bristles with fine hairs between. Flowers 2–5mm long-stalked, corolla 12–15mm long, pale yellow, later tinged reddish-brown (Greece, Crete, NW Anatolia).

Greek Golden Drop
Onosma graeca

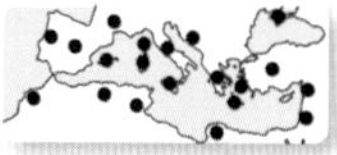

Greater Honeywort

Cerinthe major · Boraginaceae | Feb–Jun

The plant's English name derives from the old idea that this plant provided the bees with most of their honey.

Identification 0.15–0.6m. Leaves of this annual, with bristly cilia at the edge, otherwise hairless, the lower short-stalked, spatula-shaped, the upper stalkless, heart-shaped clasping the stem. Flowers in cymes, the egg-shaped bracts often tinged magenta, as long as or longer than the calyx. Corolla 15–30mm, with straight tube and short, pointed, sharply recurved lobes, more than twice as long as the calyx. The 4 nutlets joined in pairs.
Habitat Cultivated land, waste ground, roadsides.
Other species The Blue Honeywort has only 10–15mm long, somewhat upward curved, yellow, narrowed flowers which are violet at the ends (Balkan Peninsula, W Anatolia).

Blue Honeywort
Cerinthe retorta

A characteristic of Madder species is the different kinds of hairiness.

The flowers are yellow, often reddish-brown at the base or overall bluish violet.

Tree Tobacco

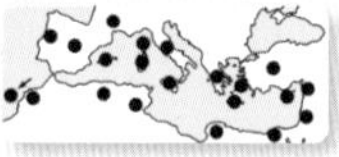

Nicotiana glauca · Solanaceae | Apr–Oct

Although this species contains less nicotine than the tobacco used for smoking, the occurrence of an additional alkaloid makes it indisputably a poisonous plant.

Identification 2–6m. Hairless, blue-green, sparsely branched, small tree. Leaves alternate, long-stalked, acute egg-shaped, entire. Flowers in loose terminal panicles, corolla hairy outside, greenish-yellow, tubular, with very short, bluntly 5-lobed limb, 25–45mm long.
Habitat Naturalised on roadsides, waste dumps, in dry river beds, native of Argentina.
Other species The plants used for tobacco products are annual, glandular hairy herbs, especially Tobacco. Leaves large, ovate lanceolate, stalkless. Flowers pink, with 5 pointed, spreading lobes. The cultivated species was introduced as a hybrid from South America.

Tobacco
Nicotiana tabacum

Wavy-leaved Mullein

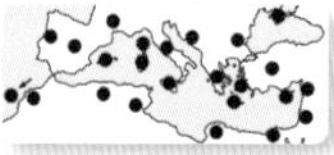

Verbascum sinuatum · Scrophulariaceae | Apr–Oct

Mulleins are represented in Europe by approx. 80 species, of which over half occur only on the Balkan Peninsula. The Turkish flora lists over 220 species.

Identification 0.5–1m. Biennial, densely and shortly grey- or yellow-felty plant. Basal leaves almost stalkless, oblong, coarsely toothed, irregularly pinnately lobed and ± wavy, especially in the non-flowering rosettes. Stem leaves with broad to heart-shaped base, stalkless. Inflorescence branched, loose, corolla 5-lobed, 15 to 30mm across, yellow, with reddish markings at base. Calyx 2–4mm long.
Habitat Roadsides, waste ground.
Other species *V. undulatum* is similar, but basal leaves stalked and so strongly wavy that the segments almost cover each other. Corolla 25–50mm across, yellow, filaments white hairy, all anthers transverse. Calyx 6–12mm long (Balkan Peninsula).

V. undulatum

The fruit capsules sit on upcurved stalks.

The purple woolly filaments with transverse anthers are conspicuous.

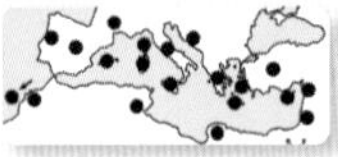

Prickly Pear fruits

Prickly Pear

Opuntia ficus-indica · Cactaceae | Apr–Jul

The Prickly Pear was introduced to Europe as a hedge plant, animal feed and source of fruit soon after the discovery of America. The species has since become naturalised in certain areas.

Identification 2–6m. Much branched, fleshy shrub with flat, reverse ovate-oblong stem segments. Small cushions of yellowish, barbed, brittle bristles, plus 0–2 strong, pale, up to 2.5cm long spines in the axils of short-lived leaflets. Flowers 5–10cm across, with numerous, yellow to orange perianth segments not divided into sepals and petals. "Prickly Pear" fruits 5–9cm, yellow to red, with inverted navel, also covered in bunches of bristles. Take care when peeling the edible, juicy fruits.
Habitat Rocky sites close to habitation, native to Central America.

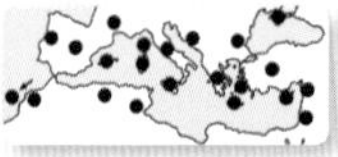

Yellow Centaury
Centaurium maritimum

Yellow-wort

Blackstonia perfoliata · Gentianaceae | May–Sep

In the Mediterranean climate, the Gentianaceae are only represented by yellow-flowering *Blackstonia* and mostly pink-flowered Centaury species. Gentians occur in the higher mountain ranges.

Identification 0.1–0.6m. Annual, hairless, upright, blue-green plant with blunt egg-shaped basal leaves. Stem leaves opposite, acute or tapering to a point ovate-triangular, joined over the entire width. Flowers yellow, 10–20mm in diameter, with short tube and 6–9 spreading lobes. Calyx deeply divided into linear, 1-veined segments.
Habitat Damp sites in maquis, on roadsides.
Other species The Yellow Centaury is the only yellow-flowering species of Centaury in the region, the flowers have only 5, rarely 6 lobes. Stem leaves stalkless, linear-ovate (Mediterranean, Canaries).

The flowers and fruits are borne on the edges of the stem sections.

In this species the stem grows through the leaves.

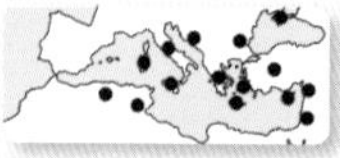

Yellow Asphodel

Asphodeline lutea · Asphodelaceae | Mar–Jun

Asphodeline species differ from Asphodelus in their more or less leafy stems. They occur in the eastern Mediterranean, especially in Turkey, where there are numerous white-flowered species in addition to a few yellow-flowered ones.

Asphodeline liburnica

Identification 0.4–1m. Perennial rhizomatous plant. Stem densely leafy to the top. Lower leaves narrowly linear, tapering to a point, triangular in cross-section, broadened into a membranous, stem-clasping sheath at base. Flowers golden yellow, in a dense, long raceme, perianth segments 2–2.5cm long, spreading into stars.
Habitat Garigue, rocky ground, also an ornamental plant.
Other species *Asphodeline liburnica* is similar, but stems only leafy in the lower half. Perianth segments 2.5–3cm long (0.25–0.6m, Jun–Jul, eastern Mediterranean, westwards to Istria and S Italy).

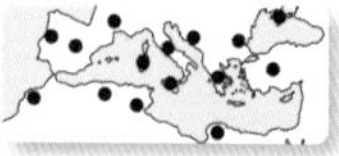

Wild Tulip

Tulipa sylvestris · Liliaceae | Apr–Jun

Finding a Wild Tulip in its native habitat is a highlight for the nature lover. Only one species is widely distributed in the Mediterranean region, but around 20 mostly red-flowering species occur locally in the eastern Mediterranean.

Tulipa saxatilis

Identification 0.1–0.45m. Perennial bulbous plant with linear-lanceolate leaves. Flowers drooping in bud, with 6 pointed, yellow perianth segments, in ssp. *australis* (photo, absent from the large islands apart from Sicily) 20–40mm long. Ssp. *sylvestris* sturdier (central Mediterranean, also Sardinia, Sicily), flowers usually larger, often greenish outside.
Habitat Ssp. *australis*: grassy areas in the montane zone, ssp. *sylvestris*: plantations, vineyards, fields.
Other species *Tulipa saxatilis* has magenta, 30–50mm long tepals, each with a yellow spot on the inside (Greece, Crete, Anatolia).

The lower of the 6 perianth segments is slightly separate.

All tepals tinged red outside, the 3 outer ones somewhat narrower.

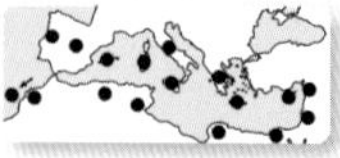

Mediterranean Asparagus
Asparagus aphyllus

Thorny Asparagus

Asparagus horridus · Asparagaceae | Apr–Nov

This asparagus well deserves its name. The short stalks (phylloclades) are modified into strong, 1–5cm long, green, usually single but also in groups of 2–3 thorns, spreading on all sides.

Identification 0.5–1m. Apparently leafless shrub with finely-grooved, grey-green stems, and with 0–2 small leaves reduced to membranous scales at the base of the phylloclades. 2–8 short-stalked flowers with 3–4mm long, yellowish, spreading perianth. Spherical, 5–8mm, bluish-black berries.
Habitat Garigue, roadsides, in dry areas.
Other species The Mediterranean Asparagus is thornless, but with 10–20mm long, distinctly unequal, spiny short shoots in bunches of 3–10(–15). Flowers 3–6, up to 5mm long-stalked, greenish-yellow. Berries also bluish-black (Sep–Oct, southern Mediterranean).

Aloe Vera
Aloe vera

Agave, Century Plant

Agave americana · Asparagaceae | Jun–Aug

The leaves of Agave are tightly appressed in the bud and after opening leave behind marks of the leaf margins. The leaves of Aloe species, in contrast, are spread in the bud and leave no marks.

Identification 3–8m. Basal rosette leaves greenish grey, fat and fleshy, 1–2m long, the edges with thorny teeth towards the top and a spiny thorn at the apex. Flower stalk with terminal panicle, flowers 6-lobed, 7–9cm long, greenish-yellow, erect at the ends of the horizontal branches. The plant dies off after the fruits are ripe.
Habitat Cultivated as an ornamental plant since the 16th century, often established in the wild. Native to Mexico.
Other species Aloe Vera (a member of the Asphodel family) has blue-green, often red-tinged leaves, thorny at the edges. Open flowers pendulous. Cultivated for the extraction of aloe vera gel (0.5–1m, native to Arabia, NE Africa).

The flower buds are tinged violet-brown.

The single flowering stem appears over a few months after 10–15 years.

Yellow Autumn Crocus, Winter Daffodil

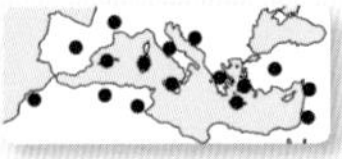

Sternbergia lutea · Amaryllidaceae | Sep–Nov

The genus *Sternbergia* is dedicated to the botanist, canon and ambassador Kaspar Graf Sternberg (1761–1838). It was and still is an honour for a genus to be named after a botanist.

Identification 0.1–0.3m. Crocus-like, autumn flowering with linear, keeled, blunt, 7–15mm wide leaves with a light mid line, appearing before or at the same time as the golden yellow flowers. The 6 perianth segments ovate-elliptical, 35–60 × 10–23mm.
Habitat Rocky ground, garigue, pasture, scrub, also an ornamental plant.
Other species *Sternbergia sicula* has only 2–6mm wide, channelled, dark green leaves with a grey mid line. Perianth segments usually narrower and acute. This species has also been placed as a subspecies of the Yellow Autumn Crocus (S Italy, Sicily, S Greece, Crete, Turkey).

Sternbergia sicula

Southern Dwarf Iris

Iris pseudopumila · Iridaceae | Mar–May

Iris species with their striking, showy flowers were named after the Greek goddess Iris who personified the rainbow.

Identification 0.1–0.3m. Sturdy rhizomatous perennial. Leaves not wider than 15mm, lance-or sickle-shaped, acute. Flowers with 5–7.5cm long, green tube and 6 roughly equally long perianth segments; the 3 inner of which erect and convergent, the 3 outer with reflexed yellow- or white-bearded lip. Flower colour blue to violet, completely yellow or with brownish lips.
Habitat Garigue, grassy areas.
Other species The Snake's-head Iris or Widow Iris has 2–24 finger-shaped nodules on the rhizome. Leaves 1.5–3mm wide, square in section. Flower bracts often exceeding the flower, the brown-violet lips not bearded (from France to Anatolia). Formerly classified in the genus *Hermodactylus*.

Snake's-head Iris or Widow Iris
Iris tuberosa

The flowers with 6 stamens have a membranous, green-edged bract at the base.

Different flower colours often occur in the same stand.

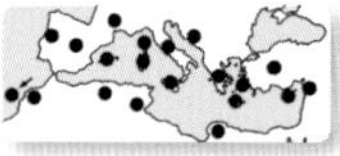

Pygmy Cudweed

Filago pygmaea · Compositae | Apr–Jun

The Pygmy Cudweed is reminiscent of the Edelweiss of the Alps (*Leontopodium alpinum*), as it also has a large mock capitulum of individual heads surrounded by spreading bracts.

F. asterisciflora

Identification 0.01–0.04m. Annual, tiny, grey felty, hairy plant, branched at base. Stem leafy, at top with a rosette of 5–16mm long, reverse egg-shaped to spatulate ± blunt, spreading leaves which surround a 5–35mm wide, dense bunch of flower heads. Involucral bracts of the individual heads numerous, brownish yellow, hairless, with 1mm long awn.
Habitat Garigue, open grassy areas, close to the coast.
Other species *F. asterisciflora* has 15–40m long, oblong-lanceolate, pointed rosette leaves below the flowering head. Involucral bracts with reflexed awn (western Mediterranean).

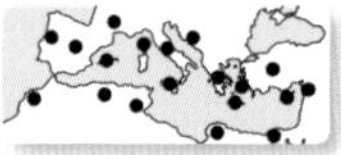

Shrubby Everlasting

Helichrysum stoechas · Compositae | Mar–Jul

Some species of *Helichrysum* have a strong aromatic scent, especially the Curry Plant which smells like its name. The young leaves and shoots are used as a culinary herb.

Curry Plant
Helichrysum italicum

Identification 0.1–0.7m. Small subshrub with linear, grey-green to white felty leaves, inrolled at the edges. Flower heads in dense corymbs with yellow disc florets and golden yellow, 4–6mm wide, spherical to broadly bell-shaped involucre. The involucral bracts membranous and without glands, arranged like loose roof tiles.
Habitat Sandy and rocky coasts, garigue.
Other species The aromatic Curry Plant has greenish, sparsely felty, rarely white-felty leaves. Flower heads with 2–4mm wide, narrow bell-shaped involucre, which is distinctly longer than broad, its bracts like flattened roof tiles, glandular (Apr–Aug, Mediterranean).

Over 30 involucral bracts surround the tiny heads of yellow disc florets.

Due to their perennial involucral bracts, *Helichrysum* species are also called “Everlastings”.

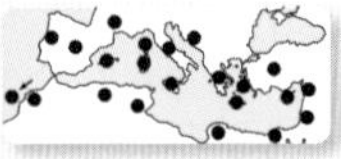

Phagnalon saxatile

Phagnalon rupestre

Phagnalon rupestre · Compositae | Mar–Jul

In contrast to *Helichrysum*, *Phagnalon* has involucral bracts with a brownish membranous margin. In the two species illustrated, the heads are always single.

Identification 0.1–0.5m. White, felty, hairy dwarf shrub. Leaves oblong-lanceolate, wavy at the edge and weakly toothed, ± inrolled, green to grey felty above, white felty beneath. Heads roughly 1cm across, with yellowish disc florets and tightly appressed, hairless involucral bracts, the middle ones blunt and smooth at the edges.
Habitat Rocky ground, garigue.
Other species *Phagnalon saxatile* has linear stem leaves with inrolled, usually smooth edges. Involucral bracts narrower and pointed, outer spreading to reflexed, middle with wavy edges (Mediterranean, absent from the east, Canaries).

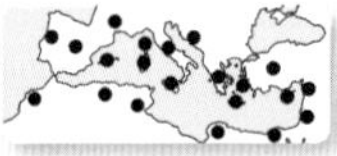

Golden Samphire
Limbarda crithmoides

Golden Samphire

Limbarda crithmoides · Compositae | Jul–Nov

***Limbarda* was previously assigned to the genus *Inula*. While it is adapted to dry sites thanks to the often dense hairiness, the hairless, fleshy leaves are an adaptation to the salt concentration at the coast.**

Identification 0.1–0.9m. Plant, woody at base with erect or ascending, little-branched, densely leafy shoots. Leaves fleshy, hairless, linear and entire or 3-toothed at apex. Small short stalks in bunches in the leaf axils. Flower heads composed of orange-yellow disc florets and yellow, 14–25mm long ray florets, projecting beyond the hemispherical involucre by twice the length. Fruitlets ribbed and hairy, with brownish-white bristly pappus.
Habitat Beaches, coastal rocks, salt marshes, with ssp. *longifolia* in the Mediterranean and ssp. *crithmoides* along the coasts of W Europe.

This *Phagnalon* species has single heads on long, leafless stems.

The flower stems are thickened and support small pointed floral bracts.

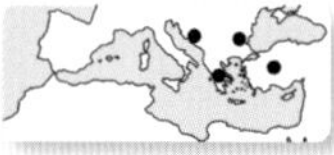

Inula aschersoniana

Inula aschersoniana · Compositae | Jun–Aug

Inula aschersoniana belongs to the *Inula verbascifolia* group, white-felty taxa in the eastern Mediterranean which are nowadays regarded as separate species and are not always easy to tell apart.

I. heterolepis

Identification 0.2–0.5m. White woolly-felty plant of rocky sites, woody at base. Lower leaves long-stalked, ovate-lanceolate, shortly wedge-shaped at base, often pointed, entire or crenate, with strongly raised veins beneath. Flower heads with yellow disc and ray florets, the latter longer than the 8–15mm wide involucre, its bracts passing over into the ± egg-shaped floral bracts. Fruitlets 2mm, with 10–15 bristles of twice the length.
Habitat Rock crevices, especially on calcareous rocks.
Other species *Inula heterolepis* is distinguished by ray florets which are shorter than the 7–10mm wide involucre (eastern Aegean, W Anatolia, Lebanon).

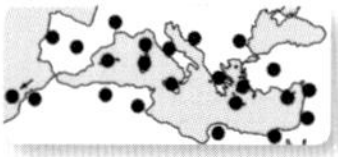

Woody Fleabane

Dittrichia viscosa · Compositae | Aug–Nov

The Woody Fleabane, previously also included in the genus *Inula*, colonises ruderal areas, often in large stands. It gives off a very aromatic, rather unpleasant smell.

Stinking Fleabane
Dittrichia graveolens

Identification 0.5–1.3m. Glandular-hairy, woody at base, densely leafy herbaceous plant. Lower leaves oblong-lanceolate, acute, usually distally toothed, upper stalkless and half clasping the stem. Flower heads with orange-yellow disc florets and yellow, 9–12mm long ray florets, distinctly projecting beyond the involucre. Fruitlets with brownish bristly pappus, joined at base.
Habitat Roadsides, waste ground.
Other species The very glandular Stinking Fleabane has a pleasant smell of camphor. Flower heads small, ray florets 4–6mm long, scarcely projecting beyond the involucre, reddish in bud (annual, 0.2–0.5m, Mediterranean).

The ray florets project beyond the involucre by 2–3mm.

The leafy, long paniculate pyramidal inflorescence supports numerous flower heads.

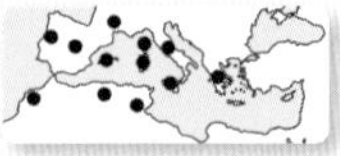

Sea Daisy

Pallenis maritima · Compositae| Mar–Jul

On rocky coastal sites the perennial Sea Daisy often forms magnificent cushions over large areas.

Asteriscus aquaticus

Identification 0.05–0.4m. Low, roughly hairy, often much branched subshrub, woody at base, with prostrate-ascending or erect branches. Leaves oblong spatulate, blunt, narrowed into stalk. Heads of golden yellow ray florets in 1 row, at 9mm as long as the outer leaf-like involucral bracts or longer. Outer fruitlets ± triangular, without wings.
Habitat Rocky ground, often close to the coast.
Other species The annual *Asteriscus aquaticus* has upper leaves which clasp the stem half way round. Heads with 5mm long, yellow ray florets in one row, 2–3 times exceeded by the outer, blunt, leaf like involucral bracts (damp, sandy sites close to the coast, Mediterranean, Canaries).

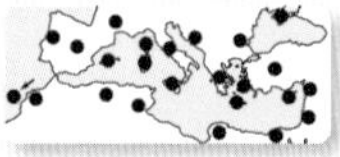

Spiny Starwort

Pallenis spinosa · Compositae | Mar–Aug

The species gets its name from the spreading, spiny involucral bracts of the flower heads.

Rose of Jericho
Pallenis hierichuntica

Identification 0.1–1m. Annual or perennial, usually hairy plant. Leaves oblong-lanceolate, shortly pointed, the upper half clasping the stem. Flowers in ± long-stalked 1.5–4.5cm wide heads, the side branches often overtopping the main stem. Outer involucral bracts spreading, like leaves and with spiny points, at 1.5–4.5cm much longer than the inner ones, and also far exceeding the 2–3 rows of 4–19mm long ray florets. Outer fruitlets compressed, winged.
Habitat Roadsides, waste ground.
Other species The Rose of Jericho is almost stemless. Involucral bracts 1.5–3.2cm long, the pale yellow ray florets in 1 row, only 3–6mm (Feb–Apr, desert plant, N Africa to Syria, Fuerteventura, local in S Spain).

The ray florets are red outside, particularly noticeable when in bud.

The flower colour of this varied species ranges from golden to pale yellow.

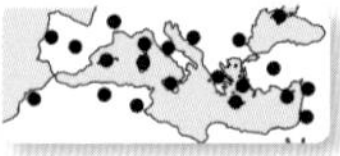

Jasonia tuberosa

Common Fleabane

Pulicaria dysenterica · Compositae | Jul–Oct

The generic name may possibly refer to the earlier use of the aromatic plant against bugs (Latin *pulex*: flea), while the species name refers to its use for dysentery.

Identification 0.2–1m. Perennial, aromatic plant with underground runners. Leaves oblong-lanceolate, hairy on both sides, especially beneath, with stalkless glands, ± wavy and distally toothed at the edge, the upper with an eared base half clasping the stem. Heads numerous, 1.5–3cm across. Pappus with toothed corona at base (as distinct from *Inula*).
Habitat Damp sites.
Other species *Jasonia tuberosa* has narrower, linear-lanceolate leaves, narrowed into the petiole. Involucral bracts spreading to reflexed (rocky sites at high altitude, SW Europe).

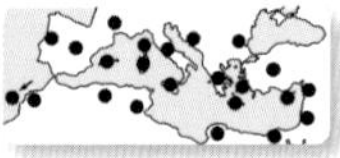

Sweet Yarrow
Achillea ageratum

Cotton Weed

Achillea maritima · Compositae | Jun–Sep

The development of a dense, white felt of hair is one possible adaptation as protection from too much evaporation, something that occurs frequently in beach plants and mountain plants.

Identification 0.1–0.5m. Aromatic, densely white felty, perennial beach plant. Stem sturdy with woody base. Leaves thick, oblong, stalkless with broad base, entire or crenate-serrate. Golden yellow disc florets in spherical, 8–10mm broad heads. This species was previously known as *Otanthus maritimus*.
Habitat Sandy beaches.
Other species The Sweet Yarrow smells of camphor and is not very hairy. Leaves with glandular dots, crenate-dentate. Flower heads 3mm across, with yellow disc and ray florets which barely project beyond the involucre, in terminal, dense corymbs (0.1–0.8m, damp ruderal areas, scrub, western Mediterranean).

The golden yellow ray florets are about 5mm longer than the hairy involucre.

The flower heads have white felty involucral bracts and only disc florets.

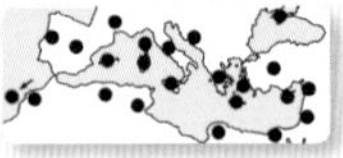

Crown Daisy

Glebionis coronaria · Compositae | Mar–Sep

Two varieties often occur together: one with only dark yellow ray florets (var. *coronaria*), the other with the ends of the ray florets pale yellow or whitish (var. *discolor*).

Identification 0.3–0.8m. Strongly scented, annual, vigorous, densely leafy, ± hairless plant. Leaves doubly pinnately parted with lance-shaped, pointed lobes, the upper clasping the stem. Flower heads single, long-stalked, 3–6cm across. Outer involucral bracts with brown, transparent membranous margin. This and the following species were previously placed with *Chrysanthemum*.
Habitat Agricultural land and waste ground.
Other species The Corn Marigold has more blue-green and fleshy, hairless, incised to coarsely toothed leaves. Flower heads with yellow disc and ray florets. Involucral bracts with narrow, brown margin (Mediterranean, Canaries, also naturalised).

Corn Marigold
Glebionis segetum

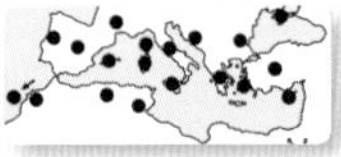

Silver Ragwort

Jacobaea maritima · Compositae | May–Oct

The species is also familiar further north in Europe and elsewhere where it is planted in gardens and on graves in several varieties under the name of Silver Ragwort or Dusty Miller as an annual, non winter-hardy, ornamental foliage plant. It was formerly known as *Senecio cineraria*.

Identification 0.25–0.8m. Subshrub with 1–2-times pinnately parted leaves, the upper side with cobwebby hairs, becoming bare, lower side densely white felty. Leaflets in west and east of Mediterranean region with different shapes. Flower heads in cymes, 10–15mm across, with golden yellow disc and paler ray florets.
Habitat Coastal rocks, also on sand.
Other species The variable annual Coastal Ragwort is fleshy. Heads 18–25mm across, involucral bracts dark pointed, numerous outer involucral bracts (0.05–0.4m, Mediterranean, Canaries).

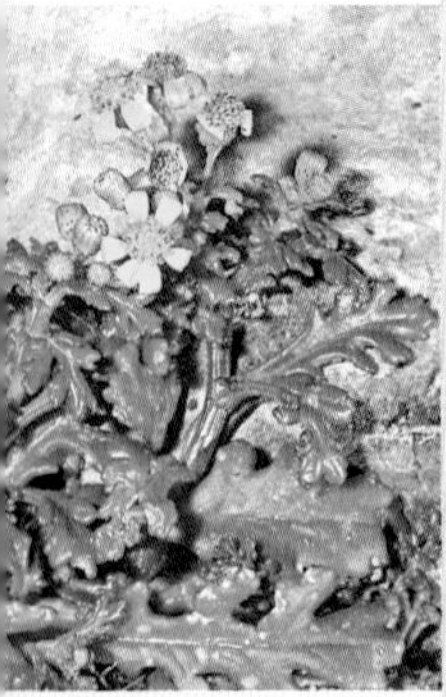

Coastal Ragwort
Senecio leucanthemifolius

Crown Daisy often forms large stands on waste ground.

The flower heads have white felty involucral bracts.

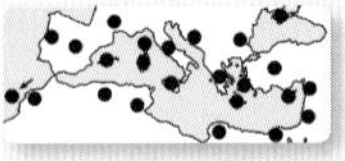

Field Marigold

Calendula arvensis · Compositae | Feb–Jun

Each flower head has 3 different fruit forms: outer hooked fruits beaked, curved, spiny, spread by animals; middle, shorter with lateral wings, wind dispersed; inner narrow, grub-shaped, "ring fruits".

Identification 0.1–0.3m. Annual, downy haired, variable species. Leaves oblong-lanceolate, somewhat wavy, entire to distally toothed, tapering to a short point, the upper half clasping the stem, stalkless. Flowers in terminal, 1–2(–3.5)cm wide heads with golden yellow disc and ray florets. Involucre 1 row.
Habitat Cultivated land, waste ground, roadsides.
Other species The Woody Marigold is woody at the base, the ray florets are more than twice as long as the involucre (western Mediterranean, Canaries).

Woody Marigold
C. suffruticosa

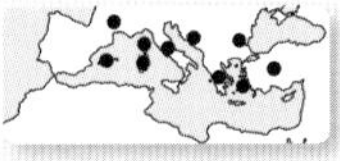

Clustered Carline Thistle

Carlina corymbosa · Compositae | Jun–Oct

The flower heads of *Carlina* have only disc florets, the inner bracts of the involucre with their dry membranous end section – which can also be silvery white or reddish in other species – simulate ray florets which expand in sunshine and close in the wet.

Identification 0.2–0.7m. Stem of the perennial, thistle-like plant branched, slightly cobwebby-felty, with spiny toothed leaves, the upper clasping the stem. Flower heads 1–2cm across, arranged like a level bunch, tongues of inner involucral bracts 10–16mm long.
Habitat Pastures, waste ground, scrub.
Other species The Acanthus-leaved Carline Thistle is stemless and has a single, stalkless, very large, 3–7cm wide flower head. Inner involucral bracts 4–6cm long (Rocky and grassy areas, mainly in mountains, S Europe).

Acanthus-leaved Carline Thistle
C. acanthifolia

The ray florets are no more than double the length of the involucre.

The Clustered Carline Thistle has golden yellow inner involucral bracts.

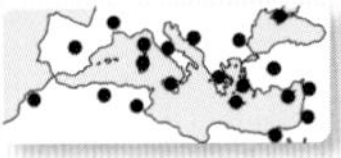

Yellow Star-thistle

Centaurea solstitialis · Compositae | Jun–Sep

Centaurea species mostly have leaves without any defence, but their involucral bracts have thorny appendages which are important for identification.

Identification 0.2–1m. Annual or biennial, grey-green, felty or woolly haired plant, stem winged from the decurrent leaves. Lower leaves lyrate-pinnately cleft, upper entire linear-lanceolate, with spiny point. Heads with 8–12mm wide, spherical involucre and pale yellow, non-glandular disc florets. Appendages of the middle involucral bracts with straw-coloured, 15–25mm long terminal thorn, with short side thorns at base.
Habitat Cultivated land, waste ground, roadsides.
Other species The similar Maltese Star-thistle has disc florets covered by sessile glands. Terminal thorn 5–10mm long, often tinged brownish-red, with side thorns up to the middle (Mar–Aug, Mediterranean, Canaries).

Maltese Star-thistle
Centaurea melitensis

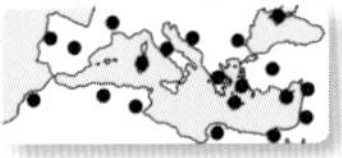

Blessed Thistle

Centaurea benedicta · Compositae | Apr–Jul

Blessed Thistle (from Latin _benedictus_: blessed) was already being used medicinally in the Middle Ages. A bitter aromatic plant, it is still used today in medicines and herbal liqueurs for stimulating the secretion of digestive juices.

Identification 0.1–0.6m. Annual, thistle-like plant. Leaves sinuately pinnately cleft, spiny toothed, with cobwebby shaggy hairs and sticky glands. Flower heads with yellow disc florets, 1.5–3cm across. Involucral bracts brown, outer thorny pointed, the middle with a pinnate thorn.
Habitat Waste ground, also established in the wild from cultivation.
Other species The perennial Hill Knapweed has rough, 1–2-times pinnate leaves. Outer flowers of the 13–17mm wide head elongated. Brown, decurrent, involucral bract appendages with long cilia and fine, protruding thorn (Jun–Aug, SW Europe).

Hill Knapweed
Centaurea collina

The involucral bract appendage ends in a long thorn.

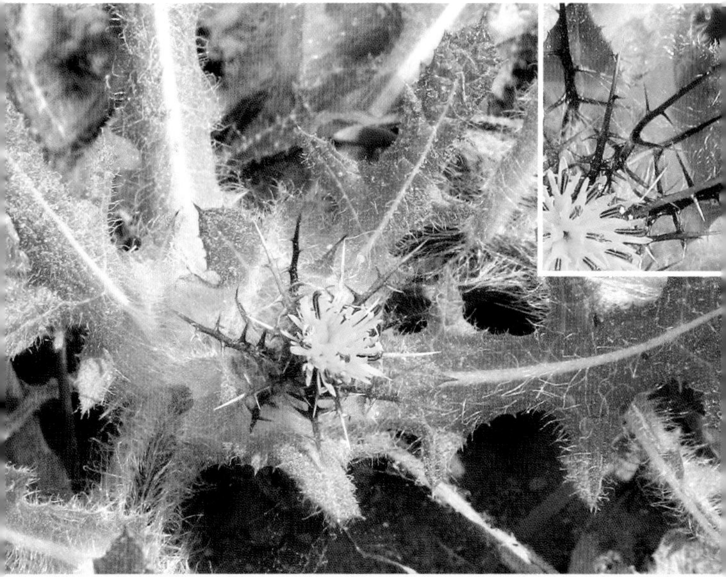

The pinnately branched thorn of the middle involucral bract is characteristic.

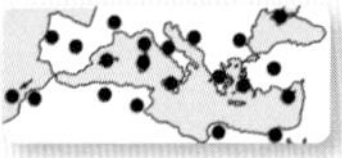

Downy Safflower, Woolly Distaff Thistle

Carthamus lanatus · Compositae | May–Oct

Safflower species owe their name to their best-known representative, the Safflower, whose flower has been used since ancient times as a substitute for true saffron (from *Crocus sativus*) for colouring foods yellow.

Identification 0.3–1m. Annual, at first cobwebby-woolly, becoming hairless, glandular, thistle-like plant. Leaves ovate-lanceolate, thorny pinnately cleft to sinuately toothed. Flower heads 2–3cm across, with yellow disc florets. Outer involucral bracts leaf-like, erect-spreading, scarcely 1.5-times as long as the middle ones; in the Smooth Distaff Thistle *Cathamus creticus* curved spreading, roughly twice as long.
Habitat Waste ground and pasture, roadsides.
Other species The Safflower has entire or finely thorny leaves. Flowers yellow to orange, as a replacement for the expensive saffron, the fruits for extraction of safflower oil (established in the wild from cultivation).

Safflower
Carthamus tinctorius

Golden Thistle, Spanish Oyster Plant

Scolymus hispanicus · Compositae | May–Sep

***Scolymus* are the only thistle-like species in the daisy family whose heads have only ray florets and which always have milky sap.**

Identification 0.2–0.8m. Perennial plant, branched from the base. Stem leaves stiff, with sinuate-thorny teeth, the margin scarcely thickened, on the stem only decurrent to the point that the stem appears to be completely winged. Flower head 1–2cm across, with golden yellow ray florets, exceeded by 1–3 spreading, thorny toothed bracts. Involucral bracts gradually tapering to a point.
Habitat Roadsides, ruderal areas, waste ground.
Other species The Spotted Golden Thistle is an annual, the stem entirely winged. Leaves and wings with very thickened, white margin, the uppermost leaves with very regular comb-like thorns. Flowers with black hairs (Apr–Jul, Mediterranean, Canaries).

Spotted Golden Thistle
Scolymus maculatus

The heads of the Smooth Distaff Thistle (top right) remain shaggy and woolly.

The thorny toothed bracts extend beyond the flowers.

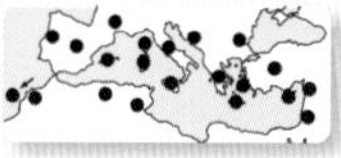

Tolpis umbellata

Tolpis umbellata · Compositae | Apr–Jun

Tolpis species can be identified by the outer bristly involucral bracts that surround the heads and continue on the stem as floral bracts.

Identification 0.1–0.3m. Annual, delicate plant with oblong spatulate, simple or toothed leaves in a rosette, and 1–2 stem leaves. Flower heads 11–16mm across, with pale yellow ray florets. Outer involucral bracts erect-spreading, roughly as long as the inner. Inner fruitlets with 4(–5) bristles.
Habitat Dry, rocky pasture.
Other species *Tolpis barbata* is much sturdier and has undivided to pinnatisect leaves. Flower heads 17–30mm across, inside with purple, outside with yellow ray florets. Outer involucral bracts bent, the inner far exceeding the outer. Fruitlets with 2 bristles (0.1–0.9m, western Mediterranean, Canaries).

Tolpis barbata

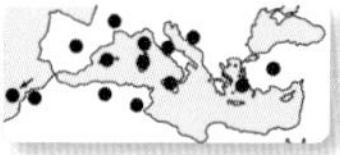

Hyoseris radiata

Hyoseris radiata · Compositae | Feb–Oct

The young leaves and flower buds of this plant are a popular addition to wild salads or soups in Mediterranean cuisine. Domestic animals like pigs, rabbits and hens also like to eat them.

Identification 0.15–0.4m. Perennial, Dandelion-like rosette plant with regularly retroserrate leaves, the sections triangular and ± toothed. Flower heads on leafless, up to 35cm long, unbranched stalks, of ray florets only, the edge ones 15–17mm long and radiating. Involucral bracts in 2 rows, inner much longer than the outer, spreading at fruiting time.
Habitat Waste ground, rock crevices, walls.
Other species *Hyoseris taurina* is woody at the base and has fleshy leaves whose right-angled sections often overlap. Flowering stems only about 15cm long (Italy including Sardinia and Sicily, Malta, Tunisia).

Hyoseris taurina

Tolpis umbellata was previously considered a subspecies of *Tolpis barbata*.

The outer ray florets, which are green underneath, are the longest and radiating.

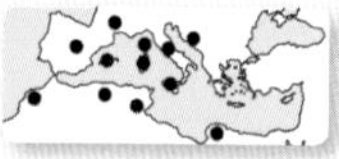

Smooth Goldenfleece

Urospermum dalechampii · Compositae | Apr–Aug

The generic name derives from Greek and refers to the long-tailed fruitlets, the English name, on the other hand, refers to the colour of the flower heads.

Identification 0.2–0.4m. Perennial, softly hairy plant. Rosette leaves sinuate-pinnately cleft, uppermost stem leaves entire and opposite. Flower heads up to 5cm across, the ray florets brownish-violet at the apex, the peripheral ones often this colour underneath as well. Fruitlets with 9–14mm long, rather oblique beak, pappus of feathery, pale rusty brown hairs.
Habitat Roadsides, cultivated land, waste ground.
Other species The annual Prickly Goldenfleece has ± deeply divided leaves with fine thorny teeth. Heads with completely yellow ray florets and spreading bristly hairy involucral bracts. Beak of fruitlets 6–8mm long, pappus white (Mediterranean, Canaries).

Prickly Goldenfleece
U. picroides

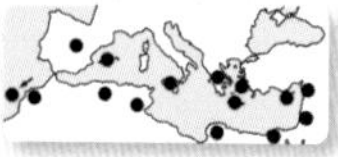

Reichardia tingitana

Reichardia tingitana · Compositae | Mar–Jun

Reichardias have several rows of involucral bracts arranged like roof tiles. These are mostly membranous, with a small tooth-like appendage at the apex.

Identification 0.05–0.4m. Usually annual plant. Basal leaves hairless, smooth to densely white pappilose, pinnately cleft, with thorny teeth at the edge, narrowed into a short, broadly winged stalk. Heads at the end of slightly swollen stalk, 2–2.5cm across. Ray florets yellow, purple at base, the marginal ones with brownish-red stripes beneath. Involucral bracts hairless, in several rows like roof tiles, with membranous, white margin up to 1.5mm wide.
Habitat On sand and rocks near the coast.
Other species The perennial *Reichardia picroides* has yellow ray florets, those on the margin have a grey-green stripe beneath. Membranous margin of involucral bracts not greater than 0.3mm wide (Mediterranean, Canaries).

Reichardia picroides

The 7–8 involucral bracts are in a row and joined at the base.

The ray florets are twice the length of the involucre.

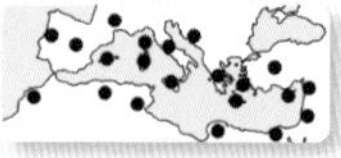

Bean Trefoil

Anagyris foetida · Leguminosae | Dec–May

The Bean Trefoil is a poisonous plant. It contains similar alkaloids to the Laburnum (*Laburnum anagyroides*). The wood was used in the past to make spears and arrowheads.

Bean Trefoil fruits

Identification 1–4m. Evil-smelling, thornless, deciduous shrub or small tree. Leaves reminiscent of *Laburnum*, stalked, trifoliate, the leaflets ovate-lanceolate with small spiny tips, silkily hairy beneath. Flowers hairless, in clusters of 3–11, corolla greenish-yellow, 18–25mm long, standard often purple at end, with blackish dots beneath, much shorter than the wings and keel. Calyx silkily hairy, bell-shaped, 2-lipped, with 5 broad teeth. Pods 6–20cm long, flat and hairless, tapering to a point at both ends, constricted between the seeds.
Habitat Rocky slopes, maquis, roadsides, usually close to the sea. Probably escaped from cultivation in places.

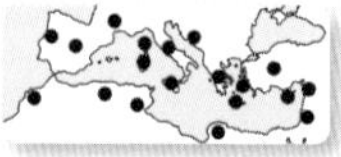

Hairy Thorny Broom

Calicotome villosa · Leguminosae | Jan–Jun

The translation of the generic name is "cut calyx" from the Greek: the upper part of the calyx separates from the lower part after flowering and initially remains like a little hat on the corolla. This distinguishes *Calicotome* from other yellow-flowered thorny shrubs.

Spiny Broom
Calicotome spinosa

Identification 0.5–3m. Squarrosely branched thorny shrub. Young twigs, underside of leaves and calyxes ± densely hairy. Leaves stalked, trifoliate, leaflets reverse egg-shaped. The golden yellow flowers with 12–15mm long standard, in clusters of 3–5 or single on branches ending in thorns. Pods 2–4cm long, densely hairy.
Habitat Garigue, maquis, locally forming large dominant stands on generally acid soils after clear-felling or fire.
Other species The Spiny Broom is similar, but mature branches and pods hairless, the flowers mostly single (SW Europe, eastwards to Italy).

The Bean Trefoil often flowers very early in the year.

The fruits are densely covered with flattened and spreading hairs.

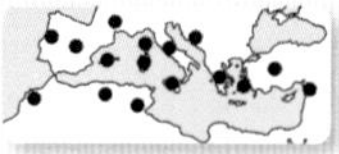

French Broom, Montpellier Broom

Genista monspessulana · Leguminosae | Jan–May

Numerous *Genista* species are thornless. Nowadays they also include the species formerly known as *Teline*. The 2-lipped calyx with deeply 2-cleft upper lip and 3-toothed lower lip is characteristic.

Identification 1–3m. Leaves stalked, trifoliate, leaflets reverse egg-shaped, hairy on both sides or only beneath. Flowers in clusters of 3–9 in the leaf axils. Calyx strongly hairy, the yellow pea flower long closed and soon falling, with 13–16mm long, hairless standard. Very hairy, 1.5–3cm long fruits.
Habitat Light evergreen and deciduous woods, maquis.
Other species The Needle-leaved Broom has almost stalkless leaves with narrow leaflets, inrolled at the edges and hairy beneath. Flowers with downy haired standard in terminal raceme (western Mediterranean).

Needle-leaved Broom
Genista linifolia

Hairy Broom

Genista hirsuta · Leguminosae | Mar–Jun

Over half of *Genista* species in the Mediterranean region are armed with thorns and are therefore protected from the grazing which has been widespread since ancient times. The majority occur on the Iberian Peninsula.

Identification 0.3–1.5m. The key characteristics are the erect-spreading, up to 3.5cm long thorns in the axils of simple, stalkless leaves with long hairs at the edges and underneath. Flowers in dense terminal racemes, standard egg-shaped, acute, 7–14mm long, ± uniformly hairy, much shorter than the long projecting keel.
Habitat Garigue, predominantly on acidic rocks.
Other species *Genista acanthoclada* has ± opposite branches ending in sturdy thorns. Leaves trifoliate with flattened hairy, narrow leaflets, the upper often simple. Flowers with 6–10mm long standard, this shorter than the keel (eastern Mediterranean).

Genista acanthoclada

Only a few flowers are open at the same time.

Every branch tip has a dense, conical inflorescence.

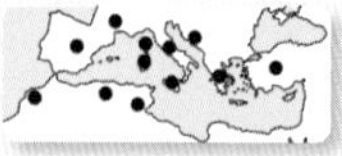

Hairy Broom

Cytisus villosus · Leguminosae | Mar–May(–Jul)

The calyxes of *Cytisus* species are bell-shaped and hairy, 2-lipped, the upper lip shortly 2-toothed, the roughly equally long lower lip 3-toothed. The species often contains alkaloids and is poisonous.

Identification 1–2.5m. Thornless, hairy shrub. Leaves stalked, trifoliate, leaflets reverse egg-shaped, the middle one longer. Flowers 2–4 on hairy stalks in the upper leaf axils, forming a leafy, racemose inflorescence. Petals yellow, standard 17–22mm long. Pods with long hairs, becoming hairless, 2–4.5cm.
Habitat Maquis, woods, on acidic substrates.
Other species The Sessile Broom is easily recognised from the ± hairless, vivid green, trifoliate leaves borne on the flowering shoots. Flowers in leafless, short, terminal racemes (woods, thickets, on limestone, up to the montane zone, SW Europe, also an ornamental plant).

Sessile Broom
Cytisophyllum sessilifolium

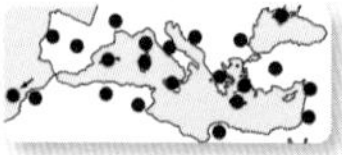

Spanish Broom

Spartium junceum · Leguminosae | Apr–Aug

This attractive species, often planted along roadsides, is considered to be a poisonous plant due to alkaloids such as cytisine. The flowers, once commonly used as a medicine, sometimes caused poisoning as a result of improper use.

Identification 1–3(–5)m. Tall shrub with rounded, hairless, erect and flexible branches. Leaves few, simple, linear-lanceolate, becoming hairless and falling early. Flowers in terminal racemes, corolla yellow, strongly scented, with 2–3cm long, pointed standard. Calyx membranous, incised above, only shortly 5-toothed.
Habitat Garigue, maquis, predominately on limestone.
Other species The smaller Common Broom has 5-angled, whip-like twigs with simple or trifoliate leaves. Flowers with 1.5–2.5cm long standard, single in the leaf axils (calcifuge, widespread in Europe, Canaries).

Common Broom
Cytisus scoparius

The standard of the pea flower has narrow, reddish-brown streaks on the inside.

The 4–12cm long, erect fruits become hairless and shiny.

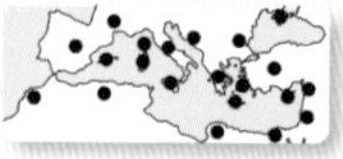

Black-eyed Vetch
Vicia melanops

Hairy Yellow-vetch

Vicia hybrida · Leguminosae | Apr–Jun

Vetches, *Vicia*, and vetchlings, *Lathyrus*, are close relatives and can be difficult to distinguish. The former generally have angled, never winged, stems and the leaves often numerous pairs of leaflets.

Identification 0.2–0.6m. Annual vetch, leaves hairy, with 3–8 pairs of leaflets and branched tendrils. Flowers short-stalked, single in the leaf axils, with 18 to 25mm long, pale yellow standard with fine purple veins, densely silkily hairy on the back. Calyx with oblique margin and unevenly long teeth.
Habitat Cultivated land, waste ground, roadsides.
Other species The Black-eyed Vetch has unusual colouring: the corolla with 17–21mm long, greenish-yellow standard, the wings blackish at the ends, the keel purple. Calyx often purple tinged, saccate at base (damp ruderal areas, S Europe, Asia Minor).

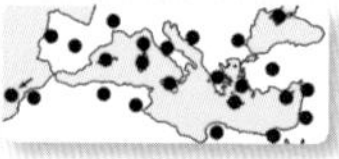

Winged Vetchling
Lathyrus ochrus

Yellow Vetchling

Lathyrus aphaca · Leguminosae | Apr–Jul

In the two yellow-flowered vetchlings illustrated, the stipules or expanded leafstalks largely take on the function of the leaves.

Identification 0.1–0.8m. Annual, green or blueish-green, hairless, often climbing plant with thin, 4-angled, but wingless stems. Leaves only developed into simple or branched tendrils, with 2 leaf-like stipules at base. Flowers mostly single, long-stalked, with yellow, 6–18mm long corolla and 5 roughly equal calyx teeth. Pods slightly bent, flat and smooth.
Habitat Cultivated land, waste ground, roadsides.
Other species The Winged Vetchling has leaf-like broadened leafstalks with only tendrils or the upper also with 1–2 pairs of egg-shaped leaflets. Flowers with pale yellow, 16–18mm long corolla, calyx teeth uneven. Pods with 2 wings on the dorsal seam (Mediterranean and Canaries).

The tendrils at the ends of the leaves are branched.

The Yellow Vetchling can be recognised by its broad, spear-shaped stipules.

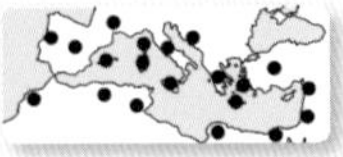

Yellow Restharrow

Ononis natrix · Leguminosae | Apr–Jul

This very variable, widely distributed and abundant species with several subspecies is glandular hairy on all parts and therefore very sticky.

Sand Restharrow
Ononis variegata

Identification 0.2–1m. Perennial, in lower part ± woody, erect, branched plant. Middle leaves trifoliate, leaflets very varied, egg-shaped to linear, mostly toothed. Flowers single in leafy, racemose inflorescences. Corolla yellow with red veins, 6–25mm long. Cylindrical, hairy, hanging pods.
Habitat Waste ground, ruderal sites, garigue, sandy coasts, predominantly on limestone.
Other species The Sand Restharrow grows ± spreading prostrate on sandy beaches. Leaves glandular-sticky, almost all simple, reverse egg-shaped, folded, sharply toothed and with distinctly raised veins. Flowers single, 10–15mm long (Mediterranean, Canaries).

Sickle-fruited Fenugreek

Trigonella balansae · Leguminosae | Mar–Jun

With its penetrating, characteristic smell, Fenugreek often makes its presence known from afar. Several species are used as culinary herbs or medicinal plants.

Furrowed Melilot
Melilotus sulcatus

Identification 0.1–0.5m. Annual, ± hairless plant. Leaves trifoliate, toothed like the stipules. Flowers in heads of 6–20 on a long stalk, corolla bright yellow, 5–8mm, the wings as long as the keel. Calyx with unevenly long teeth. Pods, curved sickle-shaped, 1.3 to 2.3cm long, compressed, hairless.
Habitat Cultivated land, ruderal areas, garigue.
Other species The Furrowed Melilot also belongs to the annuals with trifoliate leaves. Flowers 3–4mm long, in racemes of 8–28, at fruiting time as long or longer than the bracts. Pods 3–4mm, spherical, concentrically ribbed (Mediterranean, Canaries).

The flowers are single, their stalks jointed with a long awn.

The sickle-shaped, upcurved fruits are typical of this species.

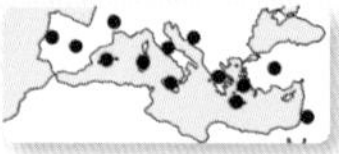

Tree Medick, Moon Trefoil

Medicago arborea · Leguminosae | Mar–Aug

Of the over 50 species in the Mediterranean area, only 3 are shrubby, few are perennials, the majority are annuals. Tree Medick has unusual snail-shaped seed pods.

Identification 1–4m. Sturdy shrub, young twigs and underside of leaves silkily hairy, leaflets entire or toothed at the apex. Flowers in heads of 8–20, corolla 12–25mm long, golden yellow. Fruits flat and spineless, 9–16mm wide, with a hole in the middle.
Habitat Rocky coasts, often also an ornamental plant.
Other species The annual Button Medick has ± hairless stems. Leaflets in upper part denticulate. Flowers 1–5, with 2–5mm long, yellow corolla. Fruits light green, later often purple, disc shaped with 4–7 loose spiral coils, spineless, 10–17mm across (0.1 to 0.6m, waste ground, roadsides, Mediterranean, Canaries).

Button Medick
Medicago orbicularis

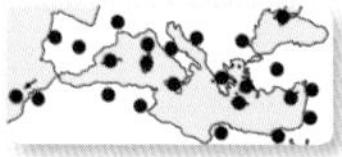

Sea Medick

Medicago marina · Leguminosae | Feb–Jun

There is a great deal of variety in the snail-shaped fruits. The size, number of spiral coils, prickles or structure are important identification features. Like Clover, *Trifolium*, all species have ternate leaves.

Identification 0.1–0.5m. Perennial plant, woody at base, prostrate, densely silvery-white hairy on all parts. Flowers in heads of 5–15, with 7–9mm long corolla. Fruits 4–7mm across, with 2–4 spiral coils, leaving a little hole in the middle.
Habitat Sand and shingle beaches, dunes.
Other species The annual Toothed Medick is ± hairless. Flowers (2–) 7–10 with yellow, 2–5mm long corolla. Fruits 4–10mm across, with 2–7.5 ± open spiral coils. 1 parallel vein on each side of the back seam, from which the often hooked curved prickles arise, rarely absent (cultivated and waste ground, Mediterranean, Canaries).

Toothed Medick
Medicago polymorpha

The fruits are spirally coiled, one and a half times at most.

The fruits have 2 rows of short spines which protrude from the felty hair.

Cretan Bird's-foot Trefoil

Lotus creticus · Leguminosae | Jan–Dec

Like some other plants named *creticus*, the Cretan Bird's-foot Trefoil does not occur on Crete, while the similar, often confused, Grey Bird's-foot Trefoil *Lotus cytisoides* occurs on the rocky coasts of Crete and throughout the Mediterranean.

Identification 0.2–1.5m long. Herbaceous perennial, densely silvery hairy Trefoil. Leaves with 5 simple leaflets, the two lower ones much longer than the midrib. Flowers 1–7, on a long stalk, corolla yellow, 12–18mm long, keel with ± straight, purple beak. Lateral calyx teeth acute, slightly shorter than the others.
Habitat Predominantly on sandy beaches.
Other species The annual Southern Bird's-foot Trefoil can be identified by the hanging, linear, flat pods curved upwards at the ends (0.1–0.5, Apr–Jun, ruderal areas, garigue, Mediterranean, Canaries).

Southern Bird's-foot Trefoil
L. ornithopodioides

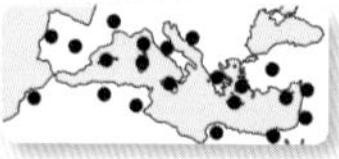

Edible Bird's-foot Trefoil

Lotus edulis · Leguminosae | Feb–Jun

The species is easy to recognise from its unusually fat pods which can be eaten when they are young and succulent.

Identification 0.1–0.5m. Annual, prostrate to ascending, sparsely hairy Trefoil. Flowers 1–3 with a trifoliate, stalkless bract, their joint stalk 2–3 times longer than the accompanying leaf. Corolla 10–16mm long, yellow to orange with reddish veined standard. All calyx teeth roughly equal, longer than the tube. Pods 1.5–3cm long and 4–8mm wide, curved upwards and pointed at the ends, furrowed beneath.
Habitat Agricultural land and waste ground, grassy areas.
Other species The distinguishing feature of the Disc Trefoil is the flat, 1–2cm wide pod with a winged, often toothed margin. Stem with soft spreading hairs, lower leaves entire, upper with 2–4 pairs of leaflets and a larger terminal leaflet (Mediterranean).

Disc Trefoil
Hymenocarpos circinnatus

The cylindrical pods often spread out on the sand like stars.

The two lower of the 5-parted leaves are distinctly separate from the other three.

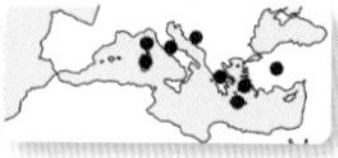

Shrubby Kidney Vetch

Anthyllis hermanniae · Leguminosae | Apr–Jul

The Shrubby Kidney Vetch is similar to some of the thorn-bearing broom species. They can be distinguished by their calyxes, the former has 5 short, regular teeth, the latter a 2-lipped calyx.

Identification 0.1–0.8m. Low, squarrosely branched shrub with twisted branches. Leaves simple or trifoliate with narrow, often folded leaflets, the middle one elongated. Flowers 1–5 in the axils of bracts, with 6–9mm long, yellow corolla, forming an interrupted inflorescence. Calyx 3–5mm, silkily hairy. Fruit egg-shaped, roughly 4mm, single-seeded, hairless.
Habitat Rocky ground, garigue, open woods.
Other species The Broom-like Kidney Vetch, a subshrub with felty whip-like branches, is thornless. Middle leaves mostly trifoliate. Flowers 1–3, corolla 9–14mm long, pale yellow. Calyx shaggy (western Mediterranean).

Broom-like Kidney Vetch
Anthyllis cytisoides

Bladder Vetch

Tripodion tetraphyllum · Leguminosae | Feb–Jun

The Bladder Vetch has changed its scientific name several times: formerly called *Anthyllis* or *Physanthyllis*.

Identification 0.1–0.5m. Annual to perennial, herbaceous, hairy plant with ± prostrate stems. Leaves odd-pinnate with 5 leaflets at most, the terminal one the largest. Flowers in clusters of 4–8, corolla with pale standard, yellow wings and an often purple-tipped keel. Calyx at fruiting time bulbously swollen, uniformly 5-toothed, a little shorter than the flower.
Habitat Roadsides, waste ground, garigue, on limestone.
Other species The calyx of the Kidney Vetch has an oblique mouth with 5 unevenly long teeth. In the Mediterranean region, besides yellow subspecies, there are also red-flowering ones such as the ssp. *rubriflora* illustrated here (France to Israel).

Kidney Vetch
Anthyllis vulneraria

The older branches are leafless and end in a long spine.

The silkily hairy calyx which is swollen at fruiting time is often tinged with red.

Scorpion Senna

Emerus major · Leguminosae | Apr–Jun(–Sep)

This plant derives its name from the slender jointed pod that resembles a scorpion's tail; previously treated botanically as *Coronilla emerus* or *Hippocrepis emerus*.

Shrubby Scorpion-vetch
Coronilla valentina

Identification 1–2m. Deciduous, ± hairless shrub with green, pinnate leaves, the 5–9 leaflets roughly equal in size. Flowers yellow, 1–8 in long stalked heads, drooping, with 14–22mm long standard, claws of petals projecting (photo spp. *major*, S Europe), in ssp. *emeroides* enclosed in the calyx. Pods pendulous, cylindrical, only slightly constricted.
Habitat Thickets, woodland edges, open woods, also cultivated, the ssp. *emeroides* from Italy eastwards.
Other species The evergreen Shrubby Scorpion-vetch often has blue-green leaves with 5–9(–15) leaflets. Flowers 4–12, with 7–13mm long standard. Fruits conspicuously divided (rocks, garigue, ornamental shrub, Mediterranean, Canaries).

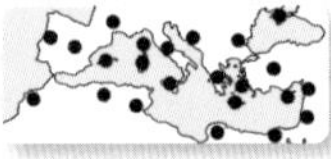

Annual Scorpion-vetch

Coronilla scorpioides · Leguminosae | Mar–Jun

Several scorpion-vetches, like those described here (except for *Emerus*) contain cardiac glycosides in the seeds and therefore have to be regarded as poisonous plants. They are of no use medicinally.

Narrow-leaved Scorpion-vetch
Coronilla juncea

Identification 0.1–0.4m. Hairless, annual crown vetch. Leaves blue-green, somewhat fleshy, simple, the upper mostly trifoliate, the rounded terminal leaf up to 4.5cm long, much larger than the side ones. Flowers pale yellow, 2–5 arranged like a crown, standard 4–6mm long, often purple veined. Pods 2–8cm, 4–6 blunt angled, strongly curved.
Habitat Cultivated land, waste ground, garigue.
Other species The up to 1m tall Narrow-leaved Scorpion-vetch is a subshrub with rush-like branches. Leaves short-lived, with 3–7 narrow, fleshy leaflets. Flowers 5–12, standard 8–11mm long. Pods 1–5cm, ± straight, bluntly 4-angled (garigue, S Europe, NW Africa).

With their long-clawed petals the flowers project far beyond the calyxes.

This species owes its name to the curved, segmented pods.

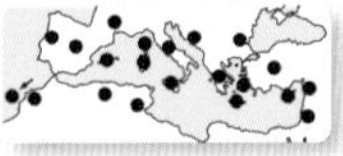

Yellow Serradella

Ornithopus compressus · Leguminosae | Feb–Jun

Leguminosae often have interesting fruits, and the two following species are examples of this.

Shrubby Horseshoe Vetch
Hippocrepis balearica

Identification 0.1–0.7m. Annual, prostrate to ascending, hairy plant. Leaves with 15–37 oblong-lanceolate, finely pointed leaflets. Flowers in heads of 1–5 on a long stalk in the axils of bracts, corolla yellow, 5–8mm long. Pods 2–5cm, ± flattened, scarcely constricted between the seeds, pendulous.
Habitat Cultivated land, pastures, ruderal areas, previously grown as cattle feed.
Other species The Shrubby Horseshoe Vetch is a small shrub (here representative of additional species, including annuals). Leaves with 12–20 narrow leaflets. The laterally compressed pods with horseshoe-shaped indentations between the seeds, as is characteristic for the genus (limestone rocks, Balearics).

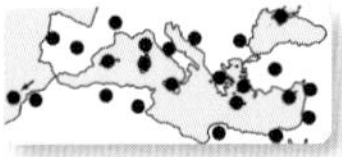

Caterpillar-plant

Scorpiurus muricatus · Leguminosae | Mar–Jun

The rolled-up fruits are somewhat reminiscent of a caterpillar. However, they often lie flat on the ground.

Scorpiurus vermiculatus

Identification 0.1–0.6m. Annual, prostrate to ascending, ± flattened hairy plant. Leaves spatula-shaped, acute, gradually narrowed towards the base, with 3–5 parallel veins. Flowers mostly 2–5, long-stalked, corolla yellow, with 5–13mm long standard. The ± irregularly spirally twisted pods often with simple, curved or bifid prickles or warts on the outer ribs. Seeds shaped like half moons.
Habitat Cultivated land, waste ground, roadsides.
Other species *Scorpiurus vermiculatus* has spreading hairs. Flowers mostly single. Outer ribs on pods with head-shaped tubercles, seeds elliptical (western Mediterranean to Italy, Canaries).

The pods with curved beak are bent downwards in a sickle-shape.

The spirally twisted pods are often covered with spines or warts.

Jerusalem Sage

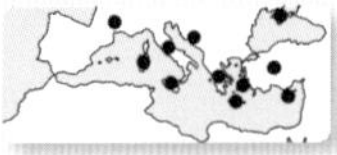

Phlomis fruticosa · Labiatae | Apr–Jul

In the Greek phrygana, several yellow-flowered *Phlomis* species often form large stands which are improved by fire and grazing.

Identification 0.5–1.5m. Dense shrub with ovate-lanceolate, simple or crenate leaves with loose hairs above and white stellate felty hairs below, the lower stalked. Inflorescence of (6–)14–36-flowered false whorls in the axils of mostly sessile bracts. Corolla 23 to 35mm long, upper lip hood-shaped, hairy outside, lower lip 3-lobed, spreading, with large middle lobe.
Habitat Garigue, open woods on limestone, ornamental shrub.
Other species The Lampwick Plant has narrower, linear-lanceolate leaves, narrowed into an indistinct stalk, woolly grey beneath. Floral bracts tapering to a point from an egg-shaped base, with 4–10-flowered false whorls (subshrub, 0.2–0.7m, Portugal, Spain, France).

Lampwick Plant
Phlomis lychnitis

White Henbane

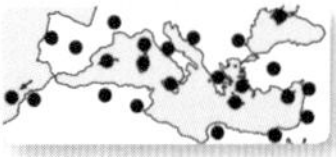

Hyoscyamus albus · Solanaceae | Mar–Sep

Henbane species are amongst the most poisonous plants. Their analgesic and intoxicating effects were known even in ancient times but nowadays they are rarely used: taking them without supervision is dangerous.

Identification 0.2–0.8m. Annual to perennial, sticky, glandular-woolly hairy plant with unpleasant smell. Leaves stalked, egg-shaped, with blunt sinuate teeth. Flowers in dense one-sided inflorescences. Corolla up to 3.5cm long, with rather unevenly 5-lobed limb, inside yellowish, the throat purple, more rarely green. Anthers scarcely projecting.
Habitat Waste disposal sites, walls, close to habitation.
Other species The Golden Henbane often grows trailing from (ancient) walls. Leaves irregularly lobed, acutely toothed. Corolla golden yellow with purple throat, anthers far protruding (eastern Mediterranean).

Golden Henbane
Hyoscyamus aureus

The hood-shaped upper lip is typical of the flowers of *Phlomis* species.

The calyx expands at fruiting time and is long-lasting.

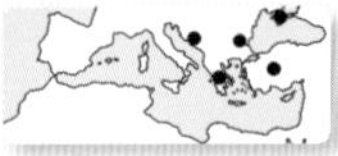

Woolly Foxglove, Grecian Foxglove

Digitalis lanata · Plantaginaceae | Jun–Aug

Foxglove species contain cardiac glycosides and are therefore poisonous. The Woolly Foxglove is the base material for the extraction of these glycosides for the production of medicines.

Identification 0.3–1m. Biennial or perennial, erect plant. Leaves oblong-lanceolate, mostly completely hairless, but the inflorescence stem and flowers glandular-woolly. Corolla 2–3cm long, yellow-brown, with a dark network of veins and white lower lip. Calyx lobes acute, without membranous margin.
Habitat Woods and thickets of the montane zone. Also planted as an ornamental and medicinal plant.
Other species The Rusty Foxglove has yellowish or reddish-brown, dark-veined flowers on a hairless stem. Middle lobe of lower lip rounded, with long hairs on both sides. Calyx lobes blunt, with broad, membranous margin (eastern Mediterranean, westwards to Italy).

Rusty Foxglove
Digitalis ferruginea

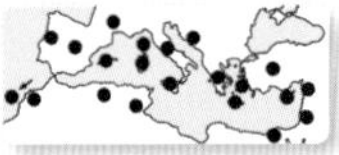

Yellow Bartsia

Parentucellia viscosa · Orobanchaceae | Apr–Sep

Parentucellia **species are hemiparasites like the genus** ***Bellardia*** **(see p. 64). The slender calyx tube of Yellow Bartsia distinguishes it from yellow-flowered forms of the Trixago Bartsia with its swollen, bell-shaped calyx.**

Identification 0.1–0.7m. Annual, erect, glandular-sticky plant. Leaves opposite stalkless, ovate-lanceolate, ± acute, crenate-serrate. Inflorescence spike-like with bracts, flowers 16–24mm long, 2-lipped, yellow, rarely white. Upper lip shortly hood-shaped, lower lip longer and much wider, 3-lobed. Calyx 10–16mm, 4-toothed.
Habitat Damp grassy areas, waste ground.
Other species The Southern Red Bartsia is an inconspicuous, glandular-sticky, often purple tinged plant max. 0.2m in height. Upper leaves almost as long as wide, incised serrate. Flowers roughly 1cm long, purple and white (Mar–Jun, Mediterranean, Canaries).

Southern Red Bartsia
P. latifolia

The white, tongue-like lower lip of the corolla has glandular hairs.

The four calyx teeth are roughly as long as the slender calyx tube.

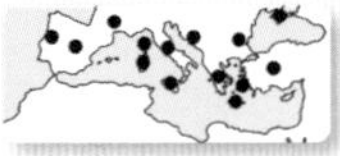

Provence Orchid

Orchis provincialis · Orchidaceae | Mar–May

Orchids owe their Greek name *orchis* not only to the testicular-shaped tubers, but also to their ancient use as an aphrodisiac. All species of orchid are now protected.

Identification 0.1–0.35m. Basal leaves lance-shaped, spotted, with 2–3 sheath-like stem leaves. The inflorescence with 5–20 pale yellow flowers and membranous bracts. Lip 8–12mm long and broad, 3-lobed, convex, with reflexed side lobes, somewhat darker in the middle with purplish-red spots, ± smooth at the edge. Spur 13–18mm, curved upwards, thickened at end.
Habitat Deciduous woods, grassy areas, maquis.
Other species The Sparse-flowered Orchid usually has unspotted leaves. Inflorescence with 3–7 flowers, lip egg-yolk yellow with fine, reddish-brown dots, denticulate at the edge, spur 14–25mm (Corsica to the Aegean).

Sparse-flowered Orchid
O. pauciflora

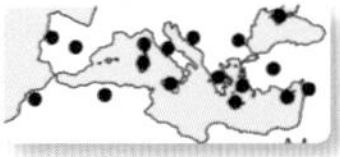

Roman Orchid

Dactylorhiza romana · Orchidaceae | Mar–May

***Dactylorhiza* species were also previously classified under *Orchis* but they differ in having finger-shaped, split tubers instead of two round ones.**

Identification 0.1–0.4m. Leaves narrow oblong, without spots. Inflorescence 4–8cm long, with yellow or light to dark violet, ± loosely erect flowers and projecting bracts. Lateral outer sepals pointing vertically upwards. Lip without markings, rounded, 9–13mm, 3-lobed with elongated middle lobe, in the red form sometimes yellow at base. Spur slender, blunt, 12–18mm, usually steeply curved upwards.
Habitat Light evergreen and deciduous woods.
Other species Barton's Orchid also has leaves without spots. Inflorescence 4–12cm long. Lip 3-lobed, with sparse red dots. Spur 9–11mm long, ± horizontal (western Mediterranean, eastwards to Italy).

Barton's Orchid
Dactylorhiza insularis

One of the few yellow-flowered *Orchis* species, most have purplish flowers.

Yellow and red-flowered plants of this species are often found together.

Sand Stock

Malcolmia littorea · Cruciferae | May–Jun

Several members of the Cruciferae with violet flowers are to be found around the coast of the Mediterranean. They include species of *Malcolmia* and *Matthiola*, which differ principally in their fruits.

Malcolmia flexuosa

Identification 0.1–0.4m. Perennial plant with stellate white felty hairs but no glandular hairs. Leaves ± stalkless, narrow oblong, entire or weakly toothed. Flowers in racemes, the 4 petals 14–22mm long, white to yellowish at base. Siliqua 3–6cm, linear, narrowed towards apex. Style with stigma soon falling, latter of 2 erect, joined lobes.
Habitat Sandy beaches.
Other species *Malcolmia flexuosa* is an annual. Leaves stalked, ovate-wedge-shaped, green and only sparsely hairy. Siliqua with persistent style (Feb–Apr, eastern Mediterranean, further established in the wild).

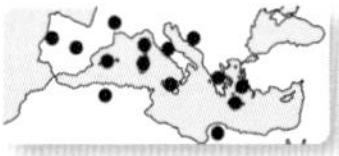

Sea Stock

Matthiola sinuata · Cruciferae | Apr–Jul

In contrast to *Malcomia* species, *Matthiola* species have stalked glandular hairs in a whitish-grey to yellowish felt of branched hairs. At the tips of the fruits are more or less prominent tubercles.

Three-horned Stock
Matthiola tricuspidata

Identification 0.1–0.6m. Biennial or perennial, beach plant, occasionally woody at base. Basal leaves sinuately toothed to pinnately cleft, with oblong, rounded lobes, upper stem leaves entire, linear-spatula-shaped. Flowers in racemes, the 4 petals 17–25mm long, whitish to violet. Siliquae 5–15cm, with only indistinct tubercles at the end.
Habitat Sandy beaches, also rocky coasts.
Other species The Three-horned Stock is an annual. Sepals with membranous margin, siliquae with 3 conspicuous, horn-like, 2–6mm long, strong, pointed, spreading processes at the ends (Mar–Jul, Mediterranean).

This *Malcomia* which is woody at the base is a decorative species of sandy beaches.

Large yellow- or black-stalked glands distinguish the hairy fruits.

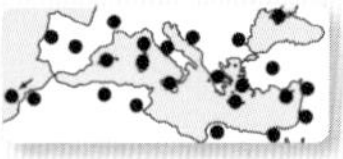

Sea Rocket

Cakile maritima · Cruciferae | all year

The Sea Rocket is a typical species of the drift-lines of sandy and shingle seashores. The fruits are buoyant and remain closed for a long time, enabling them to spread over large distances.

Identification 0.1–0.6m. Annual, prostrate to ascending, bushy branched plant. Leaves mostly stalked, hairless and often grey-green, fleshy, entire to 1–2 times pinnatisect. The scented, 4-parted flowers with 4–14mm long, violet, more rarely white petals in bractless racemes. Fruits on short, thick stalks spreading almost horizontally, 1–2.5cm long, in 2 parts.
Habitat Seashore drift-lines.
Other species The inconspicuous *Maresia nana* has entire or curved toothed leaves with stellate hairs. Petals roughly 5mm long (0.05–0.2m, Feb–Jun, sandy coasts, dune slacks, Mediterranean).

Maresia nana

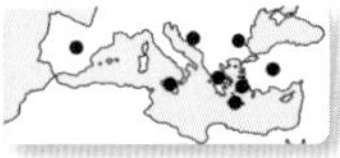

Aubrieta

Aubrieta deltoidea · Cruciferae | Apr–Jun

Aubrietas are popular rock garden plants, usually cultivars with a mixture of characters from both species described here, often with larger, sometimes double and more brightly coloured flowers.

Identification 0.05–0.2m. Perennial plant forming loose cushions. Leaves ± spatula-shaped, entire or with 1–3 denticles on each side, far exceeded by the flower heads. Petals magenta, 11–29mm long. Fruits with stellate hairs and single long, unbranched hairs, elliptical, with distinct, persistent style. Varied, like the following species.
Habitat Rock crevices, walls, elsewhere escaped from cultivation.
Other species *Aubrieta columnae* is similar but with flower heads only slightly exceeding the leaves. Fruits with only stellate hairs (Italy, northern Balkan Peninsula). The species has 4 subspecies: ssp. *italica* illustrated.

Aubrieta columnae

Two lateral projections give the fruit a spear-shaped appearance.

The flower heads project far beyond the leaves.

Poppy of Troy

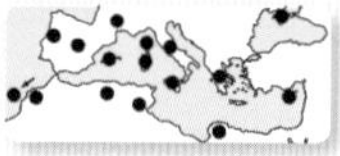

Papaver setigerum · Papaveraceae | Mar–Jul

Of the poppy species occurring in the Mediterranean region, the Common Poppy *Papaver rhoeas* – distributed throughout Europe and further afield – is the most abundant. The Poppy of Troy, which is closely related to the Opium Poppy *Papaver somniferum*, is less common.

Identification 0.4–0.7m. Annual plant with white milky sap. Leaves green, half clasping the stem, oblong-ovate, lobed-toothed, often with small bristly points. Petals 3–4cm long, white to violet, with dark spots at base. Capsule with 5–10 stigmatic rays.
Habitat Cultivated land, ruderal areas.
Other species The Rough Poppy has 1–3-times pinnatisect leaves with linear-lanceolate, awned segments. Petals an odd reddish-purple, 9–15(–25)mm long, darker at base. Capsule with yellowish, curved ascending bristles (0.2–0.4m, Mediterranean).

Rough Poppy
Papaver hybridum

Mediterranean Heath

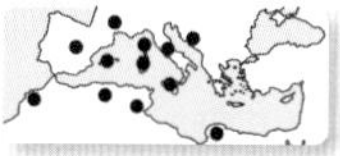

Erica multiflora · Ericaceae | Aug–Jan

The two red-flowered species of heath belong to the autumn-flowering plants of the Mediterranean, while the Tree Heath and Besom Heath flower in spring.

Identification 0.3–0.8(–2.5)m. Erect shrub with mostly hairless branches. Leaves needlelike, 6–11mm long, in whorls of 4–5, the underside covered by the inrolled margin. The terminal inflorescences dense and almost like a head, corollas pink, 4–5mm long, narrowly bell-shaped, with 4 lobes. Anthers dark red, projecting far outside the flower.
Habitat Maquis, evergreen woods, on limestone.
Other species The Whorled Heath remains a low, mostly ascending, sparsely hairy shrub. The needle-like leaves 4–8mm long, in whorls of 3–4. Flowers 1–5, lateral, corollas pink, broadly bell-shaped, 3–3.5mm long, with 4 lobes. Anthers projecting (eastern Mediterranean, westwards to Italy).

Whorled Heath
Erica manipuliflora

Sepals, stem and veins of underside of leaves have bristly hairs.

The flowers hang on long, thin, reddish stalks.

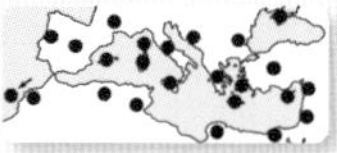

Shore Campion
Silene littorea

Small-flowered Catchfly

Silene gallica · Caryophyllaceae | Feb–Jun

***Silene* is a varied genus with numerous, often locally distributed species, of which roughly 350 occur in the Mediterranean.**

Identification 0.1–0.8m. Annual catchfly, glandular-sticky above with oblong-spatula-shaped leaves. Flowers with pink or white, ± simple, clawed petals, their limbs up to 7mm long, with small scales which form a corona. Calyx with long rough hairs, with 10 dark veins, acutely toothed. Capsule on max. 1mm long carpophore (floral axis).
Habitat Cultivated land, fallow land, pastures.
Other species The glandular hairy Shore Campion has ± fleshy, oblong leaves. Petals with pink, 5–11mm long, incised limbs, calyx bluntly toothed. Capsule on 4–9mm long carpophore (0.05 to 0.2m, beaches, Iberian Peninsula, Morocco).

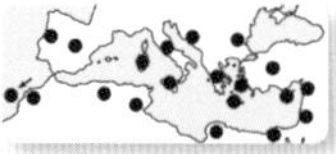

Hairy Catchfly
Silene sedoides

Cloven-petalled Campion

Silene colorata · Caryophyllaceae | Apr–Jun

A characteristic feature of Campion species is a corona of bipartite scales between the claw and limb of the petal. A longish carpophore raises the capsule in the fruiting calyx.

Identification 0.1–0.6m. Annual, finely hairy plant with opposite, spatula-shaped to linear leaves, bracts of pair usually unequal. Flowers with deep pink or white petals, their limbs 12–15mm long. Calyx with 10 hairy veins and blunt teeth, broadly club-shaped at fruiting time. Capsule on stalk of 5–7mm long carpophore.
Habitat Sandy beaches, cultivated land.
Other species The Hairy Catchfly has spreading hairs and is often red-tinged, the leaves are fleshy. Limb of petals 1–3mm long, pink, the carpophore 1–2mm (0.05–0.2m, coastal rocks, Mediterranean).

The petals of the variety *quinquevulnera* have blood-red spots (top right).

The limb of the long-clawed petals is deeply divided.

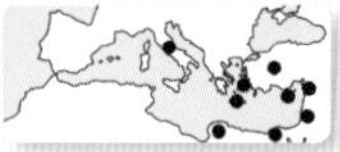

Turban Buttercup, Persian Buttercup

Ranunculus asiaticus · Ranunculaceae | Feb–May

The perianth of the Buttercup species is divided clearly into calyx and corolla. The similar anemones, on the other hand, have only a simple perianth.

Identification 0.1–0.5m. Perennial, very attractive Buttercup with flattened hairs. Basal leaves long-stalked, crenate-serrate, outer entire or 3-lobed, inner more deeply divided, stem leaves with linear segments. Flowers single, 3–6cm across, with 5(–9) wide reverse egg-shaped to wedge-shaped petals and narrow, spreading or recurved sepals. Numerous stamens with dark purple anthers. Infructescence cylindrical, up to 3cm long when ripe, fruitlets roughly 3mm, highly compressed, with sickle-shaped beak.
Habitat Rocky ground, waste ground, maquis, open woods, also grown in double varieties as an ornamental plant.

Turban Buttercup with white flowers

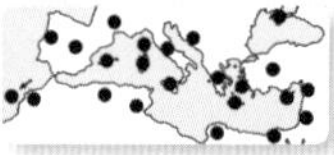

Soft Stork's-bill, Mallow-Leaved Stork's-bill

Erodium malacoides · Geraniaceae | Feb–Nov

Stork's-bills (*Erodium*) and Crane's-bills (*Geranium*) have different leaves: pinnate to pinnatisect, longer than wide in *Erodium*; rounded, palmately divided in *Geranium*.

Identification 0.1–0.6m. Annual or biennial, ± prostrate plant. Stem hairy, often glandular. Leaves heart-shaped-ovate-oblong, crenate-serrate to incised lobed. Umbel-shaped flowers 3–10, petals pink, 5–9mm long, 5 stamens. Fruit beak 1.5–3cm.
Habitat Waste ground, roadsides, grassy areas.
Other species The Corsican Stork's-bill is woody at the base and forms small, soft grey, hairy cushions. Leaves egg-shaped, the upper also lobed. Flowers mostly pink with violet veining. Fruit beak 1–1.5cm (0.1–0.2m, coastal rocks, Corsica, Sardinia).

Corsican Stork's-bill
E. corsicum

Besides red-flowered plants, yellow- or white-flowered varieties also occur.

The umbel has rounded egg-shaped floral bracts at the base.

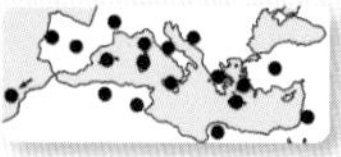

Sea Mallow
Lavatera maritima

Tree Mallow

Lavatera arborea · Malvaceae | Apr–Jun

A distinguishing feature of the flowers of the Malvaceae are the filaments joined into a tube surrounding the style. The leaves are palmately lobed.

Identification 1–3m. Biennial plant growing into a shrub or tree, with felty stellate hairs, the lower part woody. Leaves with rounded, shortly 5–7-lobed lamina. Flowers 2–7 in the leaf axils, petals 15–20mm long, magenta, veins and base darker. 5 sepals, surrounded by a trifoliate epicalyx, joined at the base, finally twice the length.
Habitat Coastal rocks, ruderal areas, ornamental plant.
Other species The Sea Mallow is a densely white-felty, stellate hairy shrub. Flowers single or in pairs with 15–30mm long, pale pink or bluish petals. The 3 outer calyx bracts almost free, somewhat shorter than the sepals (often close to the coast, western Mediterranean).

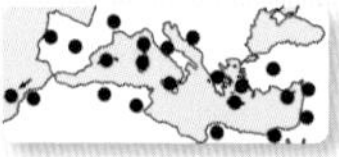

Palm-leaf Marshmallow
Althaea cannabina

Smaller Tree Mallow

Lavatera cretica · Malvaceae | Mar–Jun

Most species in the Malvaceae have a double calyx, the inner is 5-parted, the outer calyx in *Lavatera* is trifoliate, joined at the base; in *Althaea* it is divided into 6–9 segments.

Identification 0.2–2m. Annual or biennial, hairy plant. Lower leaves long-stalked, rounded, with 5–7 short, blunt, serrate-crenate lobes. Flowers 2–8, their stalks of varied length. Petals 10 to 20mm long, violet. 3 broad egg-shaped, pointed outer sepals, joined at base, shorter than the calyx.
Habitat Roadsides, waste ground, waste disposal sites.
Other species The perennial Palm-leaf Marshmallow has deeply palmately divided leaves with 3–5 narrow, blunt toothed segments. Flowers long-stalked, petals pink, 15–30mm. Outer calyx 6–9-foliate (damp sites, ditches, Jun–Sep, Southern Europe to Syria).

The petals of the Malvaceae are twisted in bud.

The pale violet petals have deeply notched tips.

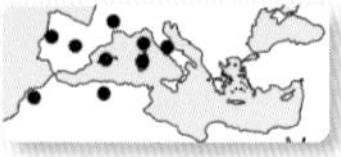

Grey-leaved Cistus

Cistus albidus · Cistaceae | Apr–Jun

The roughly 15 species of *Cistus* occur mainly in the western Mediterranean, only a third have red flowers, the remainder have white ones.

Curled-leaved Cistus
Cistus crispus

Identification 0.4–1.5m. Much branched, aromatic shrub. The opposite, stalkless, flat, egg-shaped to elliptical leaves, partially clasping the stem, with white felty stellate hairs and 3 distinctly raised veins on the underside. Flowers on 5–30mm long stalks, 4–6cm across, petals pink. Style as long as the stamens. 5 egg-shaped felty sepals.
Habitat Garigue, maquis, on limestone.
Other species The Curled-leaved Cistus is glandular hairy, its name derived from the stalkless leaves with wavy curled margins. Flowers only 1–5mm long-stalked, 3–4cm across, petals deep pink (0.3–0.6m, on acid soils, western Mediterranean).

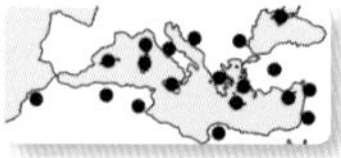

Cretan Cistus

Cistus creticus · Cistaceae | Apr–Jun

The Cretan Cistus is the most widely distributed of the pink-flowered species. It is very varied, 3 subspecies have been described.

Small-flowered Cistus
Cistus parviflorus

Identification 0.3–1.5m. Shrub with ovate-lanceolate, green to grey-green on both sides, ± densely hairy leaves, with depressed pinnate veins above, leaf margins smooth to strongly wavy. Stalks 3–15mm long, broadened at base. Flowers pink, 4–6cm across, style as long as the stamens. 5 long pointed sepals.
Habitat Garigue, maquis.
Other species The Small-flowered Cistus has smaller, ovate-elliptical leaves, with grey felty stellate hairs on both sides. Flowers 2–3cm across, petals pale pink, notched at the tip. Only red-flowered species with stalkless stigma (0.3 to 1m, Mar–May, from Lampedusa eastwards to S Anatolia and Cyprus, Libya).

The outer sepals are somewhat heart-shaped and taper sharply to a point.

The plant shown is ssp. *eriocephalus* with long, white hairs.

Spring Sowbread, Wavy Cyclamen

Cyclamen repandum · Primulaceae | Mar–May

Cyclamens are well-known (poisonous) houseplants. A number of species delight the nature lover in spring and autumn in the Mediterranean region.

Identification 0.1–0.2m. Perennial plant with large tuber. Leaves developing in spring before the flowers, broad heart-shaped, tapering to a point, with irregularly toothed to sinuately curved margin, dark green above with lighter spots. Flowers usually pink, corolla lobes 15 to 30mm long, not eared. Fruit stalk spirally coiled from the end.
Habitat In the undergrowth of shady woods.
Other species The Sowbread or Ivy-leaved Cyclamen flowers in autumn before the appearance of the heart-shaped, blunt lobed leaves with irregularly finely toothed margin. The recurved corolla lobes roughly 20mm long, eared at base and with dark purple double marking (Aug–Nov, France to western Anatolia).

Sowbread or Ivy-leaved Cyclamen
C. hederifolium

Oleander

Nerium oleander · Apocynaceae | Jun–Sep

The evergreen shrub stands out in summer with its pink, or more rarely white, flowers. Like Foxglove, the plant contains poisonous cardiac glycosides, and there have been reports of fatalities after using Oleander branches as skewers.

Identification 1–4m. Hairless shrub with milky sap. The leathery, lance-shaped leaves, narrowed into the stalk and somewhat inrolled at the edges, in whorls of 3–4 or opposite. Flowers in terminal corymbs, corolla 3–4cm across, with funnel-shaped tube and 5 obliquely truncated, spreading tips, with cleft appendages in the throat. Erect, 8–18cm long, reddish-brown fruits. Seeds with a tuft of hair.
Habitat On river banks, also in intermittently dry riverbeds. Cultivated in a range of varieties, some with double flowers, also tree-shaped.

Oleander with double flowers

The flowers have 5 recurved corolla lobes with no ears at the base.

The Oleander is a species of river banks, nowadays also planted along motorways.

Sea Bindweed

Calystegia soldanella · Convolvulaceae | Apr–Oct

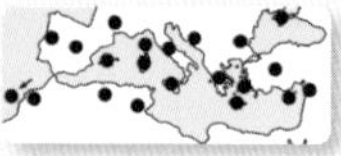

With its large flowers, the species is amongst the most attractive plants of Mediterranean coastal dunes. Its dark-green, kidney-shaped, somewhat fleshy leaves are reminiscent of the Snowbell *Soldanella*.

Identification Up to 1.5m far creeping, perennial, non-twining, hairless plant. Leaves 1–2-times as wide as long, kidney-shaped, with 3–6cm long stalk. Flowers funnel-shaped, 2.5–4cm long, pink, with 5 white stripes. A pair of wide, egg-shaped bracts encloses the calyx (difference from *Convolvulus*).
Habitat Coastal dunes, Mediterranean coasts and beyond.
Other species The Large Bindweed, in contrast, is a hairless climbing plant with thin, arrowhead-shaped leaves. Flowers with 5–9cm long, white corolla. The 2 overlapping bracts are swollen, saccate, covering the sepals (0.5–3m, thickets, Mediterranean).

Large Bindweed
Calystegia silvatica

Mallow-leaved Bindweed

Convolvulus althaeoides · Convolvulaceae | Mar–Jun

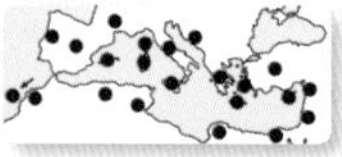

This is the most widely distributed Bindweed in the Mediterranean area. Its leaves are very varied, from entire to sinuately lobed. The closely related Narrow-leaf Bindweed, *C. elegantissimus*, commoner in the east, has silvery flattened hairy leaf segments.

Identification 0.3–1m. Perennial, spreading hairy plant with prostrate or twining stems. Leaves stalked, heart- to arrowhead-shaped, crenate-lobed, ± deeply divided, with broad segments. The 2.5–4cm long, pink flowers 1–3 in the axils.
Habitat Roadsides, cultivated land, fallow fields.
Other species The Pygmy Bindweed has flattened silky hairs. Leaves narrow, folded, the lowest with a membranous extension at base. Flowers 1–7 with 12–25mm long, pale pink corolla (0.05–0.25m, Apr–Aug, rocky ground, garigue, Mediterranean).

Pygmy Bindweed
Convolvulus lineatus

Sea Bindweed has extensive runners.

The large, wide, funnel-shaped flowers are folded and twisted when in bud.

Hottentot-fig

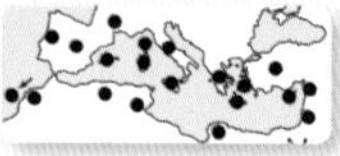

Carpobrotus edulis · Aizoaceae | Mar–Jul

Cultivated as an ornamental plant and to stabilise embankments and dunes, this species, native to S Africa, is now naturalised in the Mediterranean region and suppresses the natural vegetation in some places.

Identification Up to 2m long. Becoming woody at base, prostrate to hanging species forming dense mats. The fleshy leaves erect, acute, joined in pairs at base, triangular in cross-section. Flowers 8–12cm across, with numerous narrow, crimson or yellow petals and matching crimson or yellow filaments. The sweet and sour, slimy flesh of the fruit is edible.
Habitat Often naturalised, native to S Africa.
Other species The Baby Sun Rose has flat, fleshy, heart-shaped leaves and roughly 2cm wide flowers (Apr–Nov, naturalised in SW Europe, Crete).

Baby Sun Rose,
Heartleaf Ice Plant
Aptenia cordifolia

Peony

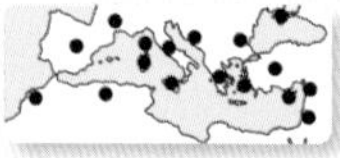

Paeonia mascula · Paeoniaceae | Mar–Jun

Over 10 species of peony grow in woods in the Mediterranean region. The Garden Peony *Paeonia officinalis* also comes from S Europe, the double-flowered form is familiar from gardens.

Identification 0.2–0.8m. Perennial herbaceous plant with large, double tripartite leaves, the lower with 9–16(–21) oblong-lanceolate to elliptical, pointed, sometimes reddish-tinged leaflets. Flowers 8–14cm across, the 5–8 petals red, more rarely white. Usually 3–5 shaggily hairy follicles. Poisonous.
Habitat Deciduous forests, thickets in montane zone.
Other species The Majorcan Peony has 3–9 leathery, egg-shaped, pointed leaflets, tinged purple beneath. Flowers 6–12cm across, with 5–8(–10) unusual pink-purple coloured petals. Usually hairless follicles (shady rocky sites, Balearics).

Majorcan Peony
Paeonia cambessedesii

Crimson- and yellow-flowered plants are now considered to be varieties of the same species.

The large, red or white impressive flowers of the Peony.

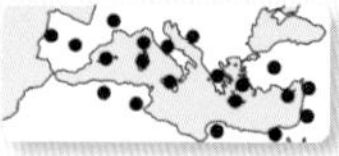

Poppy Anemone, Crown Anemone

Anemone coronaria · Ranunculaceae | Feb–Apr

This attractive Anemone can be bought in late winter as a cut flower in various colours and also with double flowers.

Peacock Anemone
Anemone pavonina

Identification 0.1–0.45m. Perennial plant with tuberous rhizome. Basal leaves long-stalked, usually trifoliate, the segments deeply lobed. Stem hairy, 1-flowered, with 3 whorled stalkless, incised bracts. Perianth segments 18–45mm long, mostly 6(–8), silkily hairy beneath, variable colours, from bright red, pink, blue and violet to white. Unlike *Ranunculus asiaticus* the species has no sepals.
Habitat Fallow fields, rocky ground, garigue.
Other species The leaves of the Peacock Anemone are less divided, the floral bracts mostly entire. Perianth segments 7–12, in red and blue shades, often paler at base, white or yellow (local from France to Turkey).

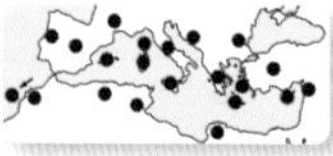

Pomegranate

Punica granatum · Lythraceae | May–Sep

From ancient times the Pomegranate has been seen as a symbol of fertility due to its large number of seeds. It also provided the names for the city of Granada, the grenade and the semiprecious stone garnet.

Pomegranate design on a Roman sarcophagus

Identification 2–7m. Hairless, often thorny shrub or tree with poisonous, alkaloid-containing bark. The deciduous leaves ± opposite, oblong-lanceolate, entire. Flowers 1–3, with 5–8 crumpled, 2–3cm long, bright red petals and numerous stamens. Fruit similar to an apple up to 9cm, the rind leathery, reddish brown. Seeds numerous, embedded in an edible, pleasantly sour-sweet tasting, juicy, jelly-like seed coat. The seed oil is frequently used nowadays in cosmetics.
Habitat Fruit and ornamental tree, sometimes with double flowers, locally naturalised. Native to SW Asia.

Plants with different flower colours often grow together in one stand.

The red flowers sit in a fleshy calyx and hypanthium of the same colour.

Rosy Garlic

Allium roseum · Amaryllidaceae | Feb–Jun

The numerous *Allium* species of the Mediterranean region have very varied leaves, from tubular to triangular to flat, as in the case of the following representatives.

Identification 0.1–0.8m. Perennial, bulb with crusty outer involucre, characteristically dotted with small holes. Leaves flat, finely toothed at the edge (hand lens needed!), up to 14mm across, sheathing the flower stalk below by up to one fifth. The hemispherical cymose umbel often with bulbils, up to 7cm across, perianth cup-shaped, ± erect on long stalks, pink or white, 7–12mm.
Habitat Cultivated land, waste ground, garigue.
Other species The vigorous, up to 1m high Black Garlic has basal, flat, up to 8cm wide leaves. Cymose umbel dense-flowered, 5–10cm across, sepals spreading like a star, pinkish-purple or white, outside with greenish central vein (Mediterranean, Canaries).

Black Garlic, Broad-leaved Leek
Allium nigrum

Fritillary

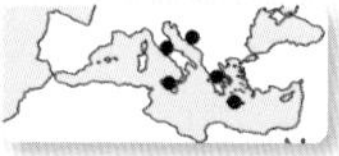

Fritillaria messanensis · Liliaceae | Mar–Apr

Fritillaries are uncommon in the Mediterranean region, most occur only locally in the eastern Mediterranean (Greece 20, Spain 3), only *Fritillaria messanensis* is more widely distributed.

Identification 0.15–0.3m. Perennial bulbous plant. Lowest leaves often opposite, linear, 3–7mm across, the others alternate or the topmost in whorls of three in ssp. *messanensis*. Flowers mostly single, drooping, broadly bell-shaped, their sepals 22–42mm long, with brownish or purplish indistinct "chess board" pattern, with broad, green strip in the middle. 4 subspecies.
Habitat Grassy areas, thickets, open woods.
Other species *Fritillaria graeca* is similar, the leaves 11–25mm wide, the lower often opposite, the uppermost single. Sepals 18–22(–30)mm long (southern part of Balkan Peninsula, Aegean, not on Crete).

Fritillaria graeca

The flowers are enclosed by a membranous, 3 or 4-lobed involucre of bracts.

Of the four subspecies, ssp. *sphaciotica* (top right) grows only on Crete.

Syrian Thistle

Notobasis syriaca · Compositae | Apr–Jun

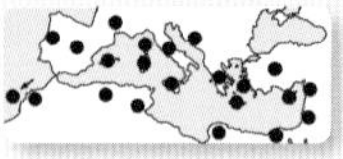

Amongst the many red-flowered thistle-like Compositae, *Notobasis* is easy to recognise: the upper, spiny, mostly magenta-tinged leaves with broad white main veins surround and extend beyond the inflorescence.

Identification 0.3–1.5m. Annual species with non-winged stems and thorny pinnatisect, white-veined leaves. Flower heads 12–25mm across, with purple disc florets. Fruitlets with long feathery hairs outside and short plain ones inside.
Habitat Roadsides, waste ground, cultivated land.
Other species The perennial Cretan Thistle has a thorny, winged stem right to the top. Leaves with inrolled margin and 5–20mm long thorns. As with all *Cirsium* species, fruitlets with feathery pappus only (Jul–Aug, damp pastures, hollows, eastern Mediterranean).

Cretan Thistle
Cirsium creticum

Purple Milk Thistle, Boar Thistle

Galactites tomentosus · Compositae | Mar–Aug

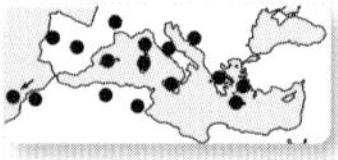

The plants known as thistles belong to various genera. However, what they have in common is the thorny armouring on leaves and stems which provides protection against being eaten by grazing animals.

Identification 0.1–1m. Annual or biennial plant. Stem leaves pinnately parted, with 1.5–6mm long thorns, white veined or mottled above, white felty beneath, somewhat decurrent on stem. Flower heads 1–1.5cm across, with only disc, pink, mauve or white flowers, the outer infertile, more brightly coloured and radiating. Involucral bracts with 5–10mm long, greenish thorn. Fruitlets with white, feathery pappus.
Habitat Roadsides, waste ground, cattle pasture.
Other species *Galactites duriaei* has winged stems and much longer thorns. Heads almost stalkless, several on one stem (SE Spain, NW Africa).

Galactites duriaei

The short-thorned involucral bracts have fine cobwebby hairs.

Purple Milk Thistle gets its name from the white markings on the leaves.

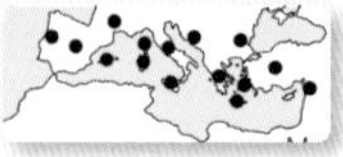

Illyrian Thistle

Onopordum illyricum · Compositae | May–Aug

The imposing *Onopordum* species is unmistakable. Their size, the thorny winged stem with massive flower heads and the often grey-white hairiness also make them decorative ornamental plants.

Identification 0.5–2m. Biennial, grey to white felty hairy plant. Stem with up to 10mm wide thorny wings. Leaves stalkless, thorny, deeply pinnately cleft. Flower heads 4–6cm across, the purple disc florets with glands. Involucral bracts pointed, with a short thorn. Pappus hairs feathery, joined at base.
Habitat Pastures, roadsides, ruderal areas.
Other species *Onopordum tauricum* is green and glandular-sticky, the stem with wings up to 15mm wide. Flowers 5.5–8cm across, with ± cobwebby hairy, often violet involucre. Its bracts glandular, narrowed into a long thorn. Pappus hairs rough (Sardinia to Syria).

Onopordum tauricum

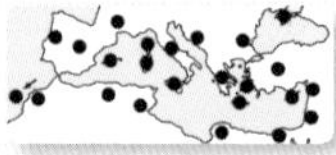

Milk Thistle

Silybum marianum · Compositae | Apr–Aug

The Milk Thistle is an old medicinal plant for treating liver and gall bladder complaints. Nowadays the silymarin isolated from the fruit peel is particularly important. It is a protective agent for the liver, used for example after acute poisoning from the Death Cap mushroom (*Amanita phalloides*).

Identification 0.3–2m. Biennial plant with large, overwintering leaf rosette. Leaves sinuately-thorny lobed, hairless or becoming so, shiny green, extensively white veined. Flowers magenta, all disc-shaped, in long-stalked, 4–8cm broad, erect or slightly nodding heads. Outer involucral bracts with spiny toothed appendages, ending in a strong, yellow, 2–5cm long, recurved thorn. Fruits 6–8mm, grey mottled, the pappus with long external and shorter internal bristles, joined at the base.
Habitat Waste disposal sites, roadsides, cattle pastures.

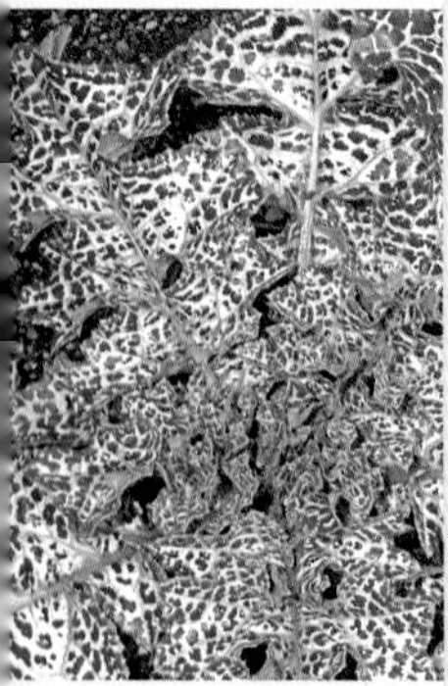

Leaf rosette of the Milk Thistle

The outer and middle involucral bracts are strongly recurved.

The fruits, which are used pharmaceutically, have a clear, yellow edge at the top.

Round-headed Knapweed

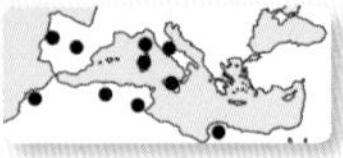

Centaurea sphaerocephala · Compositae | Apr–Jun

The Knapweeds *Centaurea* are present in the Mediterranean with around 400 species, but many only occur over very small areas.

Identification 0.05–0.7m. Perennial, prostrate to erect, variable species. Leaves roughly hairy to cobwebby-felty and sticky, lyre-shaped pinnately cleft, thorny pointed, upper toothed or entire. Heads single, made up of purple disc florets ± radiating at the edge. Involucre 12–35mm across, involucral bract appendages reddish brown, with 5–13 yellowish, palmately arranged, 3–6mm long thorns, the central one usually longer.
Habitat Sandy beaches, pasture.
Other species The biennial Red Star-thistle is squarrosely branched. Peripheral flowers scarcely expanded, involucral bracts with spreading, 10–18mm long thorn, pinnate at base (Mediterranean, Canaries).

Red Star-thistle
Centaurea calcitrapa

Centaurea raphanina

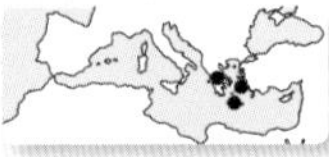

Centaurea raphanina · Compositae | Apr–Jul

The following species of sandy coasts and rocks are examples of the numerous Knapweeds in the eastern Mediterranean.

Identification 0.05(–0.2)m. Variable, ± stemless rosette plant. Flower heads 2–4 with purple disc florets, the peripheral ones scarcely radiating. The ssp. *mixta* illustrated with interrupted lyre-shaped pinnately cleft leaves, the larger leaf segments toothed (S and E Greece, Aegean). Two further subspecies with simple terminal thorn and untoothed leaf segments or scarcely pinnate leaves on Crete.
Habitat Sandy coasts, rock crevices.
Other species *Centaurea pumilio* grows on sandy coasts. Involucral bracts with finely toothed, pale margin and 5–9mm long terminal thorn, peripheral flowers radiating (Mar–Aug, eastern Mediterranean).

Centaurea pumilio

The western Mediterranean ssp. *polyacantha* has radiating peripheral flowers.

The involucral bracts of ssp. *mixta* have a 9–25mm long, pinnate terminal thorn.

Purple Salsify, Common Salsify

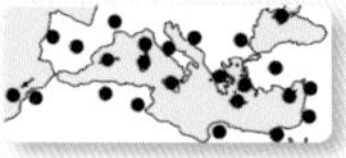

Tragopogon porrifolius · Compositae | Apr–Jul

The angled, sometimes scaly fruitlets measure 2–5cm including their long beak. The brownish pappus hairs are joined at the base, roughly half the length, feathery and intertwined with each another.

Identification 0.2–1.2m. Biennial, ± hairless plant. Leaves linear, gradually tapering to a point, often wavy, broadened at base and half clasping the stem. Flower heads on stalks thickened above, involucral bracts mostly 8. Ray florets wine-red, in ssp. *eriospermus* (photo) at least two-thirds as long as the involucral bracts, in other ssp. max. two-thirds as long.
Habitat Grassy areas, waste ground, roadsides, formerly cultivated as a vegetable and established in the wild.
Other species The annual, smaller Slender Salsify, has varying fruitlets: outer with 5 unevenly long, simple bristles, inner with feathery hairs. Involucral bracts much longer than the pink flowers (Mediterranean).

Slender Salsify
Geropogon hybridus

Pink Hawk's-beard

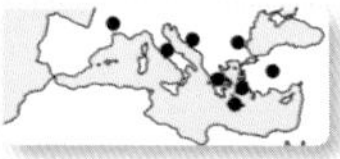

Crepis rubra · Compositae | Apr–Jun

The often difficult to identify *Crepis* species distributed throughout the Mediterranean region and further afield mostly have yellow flowers, while the pink-flowered members of the genus are easily recognisable.

Identification 0.05–0.4m. Annual plant, leaves almost all basal, simple and toothed to deeply pinnatisect with pointed lobes. Stem with few bract-like leaves and mostly one head with pink ray florets. Outer involucral bracts roughly as long as the inner glandular hairy ones. Fruitlets 10-ribbed, beaked and with a soft, pure white pappus.
Habitat Cultivated land, grassy areas.
Other species The usually yellow-flowering Beaked Hawk's-beard occurs in Sicily as the pink-flowered ssp. *bivonana*. It is distinguished by the curved, wide membranous, outer involucral bracts (Mediterranean, absent in east).

Beaked Hawk's-beard
Crepis vesicaria

The attractive fruiting head forms a sphere about 10cm across.

When fruiting, the white pappus extends beyond the glandular hairy involucre.

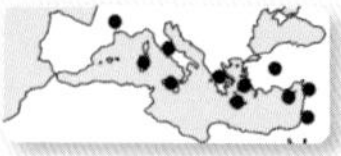

Climbing Birthwort

Aristolochia sempervirens · Aristolochiaceae | Mar–Jul

Aristolochia species have refined fly-trap flowers which keep insects for pollination trapped inside the bulbously expanded tube until fertilisation, by means of spreading hairs on the inner surface. Species with U-shaped curved, large flowers are of local occurrence, especially in the eastern Mediterranean.

Identification Climbs to 3m. Evergreen, woody plant. Leaves ± long-stalked, leathery, dark green, heart-shaped ovate-lanceolate, the top ones hairless. Flowers 2–6cm, lip rounded to egg-shaped, yellow with reddish-brown margin and purple stripes, shortly hairy inside.
Habitat Shady, damp thickets, woods.
Other species The Andalusian Dutchman's Pipe has a lip with dense short hairs inside. Leaves somewhat blue-green (Dec–Jun, Iberian Peninsula, NW Africa).

Andalusian Dutchman's Pipe
Aristolochia baetica

Spanish Birthwort

Aristolochia pistolochia · Aristolochiaceae | Apr–Jun

Other species of *Aristolochia* have erect flowers with a straight tube. After pollination the tube's opening is partially covered by the lip and the flower bent downwards. *Aristolochia* species are somewhat poisonous and in the past were used for a variety of medicinal purposes.

Identification 0.2–0.8m. Perennial, herbaceous, hairy plant. The heart-shaped leaves short-stalked, with fine, cartilaginous teeth or warts on the underside. Flowers 2–5cm long, with almost straight, brownish tube, lower half of inside of lip striped.
Habitat Dry sites, garigue, cultivated land.
Other species In the Yellow Birthwort the flower is 2.5 to 7.5cm long, the straight tube somewhat bulbously expanded in the middle section, greenish yellowish, the tongue shorter and narrower than the tube. Leaves finely hairy (deciduous woods, eastern Mediterranean, westwards to Italy).

Yellow Birthwort
Aristolochia lutea

Climbing Birthwort often climbs through and far above shrubs.

The species can be easily identified by its crinkly leaf margins.

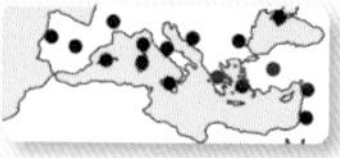

Judas Tree

Cercis siliquastrum · Caesalpiniaceae | Mar–Apr

Various stories have grown up around the Judas Tree, for example Judas is supposed to have hanged himself from a tree of this kind.

Identification 3–5(–10)m. Deciduous tree or shrub, leaves stalked, rounded, blunt or notched at the tip, heart-shaped at the base, totally hairless. Numerous pinkish, up to 2cm long flowers borne in small racemes on the older branches, similar to Leguminosae but with a different arrangement when packed in a bud: the two lower, larger petals cover the two lateral ones and these cover the upper one; in the Leguminosae it is the other way round. 10 free stamens. Calyx broadly bell-shaped. The reddish-brown pods 6–15cm long, flattened, narrowly winged on the ventral seam.
Habitat Riparian forest, maquis, rocky slopes, native in eastern Mediterranean, more widely distributed as an ornamental tree.

Fruit of the Judas Tree

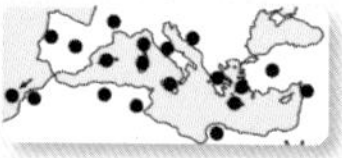

Spanish Vetchling, Crimson Pea

Lathyrus clymenum · Leguminosae | Mar–Jun

In contrast to vetches *Vicia*, vetchlings *Lathyrus* often have a winged stem and only 0–3(–5), often parallel-veined pairs of leaflets.

Identification 0.3–1m. Usually annual, hairless, climbing plant with winged stem. Lower leaves reduced to the widened leaf stalk (!), upper with 2–5 pairs of linear leaflets on winged rachis and with branched tendrils. Flowers 1–5, long-stalked, corolla multicoloured, with 12–22(–25)mm long, red or pink standard, violet or pale pink wings and whitish keel. Calyx teeth shorter than tube. Variable species.
Habitat Agricultural land and waste ground, roadsides.
Other species The Red Vetchling or Red Pea is dainty, the leaves have 1 pair of lance-shaped leaflets. Flowers single, corolla crimson, 10–14mm long. Calyx teeth 2–3-times as long as the tube (Mediterranean, Canaries).

Red Vetchling, Red Pea
Lathyrus cicera

The flowers grow directly from the older branches, often even before the leaves appear.

Standard, wings and keel have different colours.

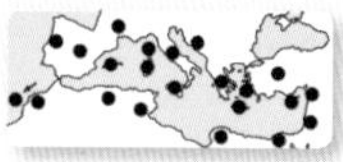

Starry Clover

Trifolium stellatum · Leguminosae | Mar–Jun

Clover species *Trifolium*, of which the Mediterranean has over 150, are herbaceous with trifoliate, mostly toothed leaves. Many have interesting calyx shapes which are important for identification.

Identification 0.05–0.4m. Annual plant with erect stems with flattened or spreading hairs. Flowers in spherical to egg-shaped, long-stalked, 15–25mm heads. Corolla 8–12mm, pink or also white, ± as long as the silkily hairy calyx. The long pointed, roughly equal calyx teeth which are twice as long as the tube are characteristic.
Habitat Cultivated land, waste ground, roadsides, garigue.
Other species The Woolly Clover, with heads reaching 15mm across, has twisted flowers, so that the keel is on top. Upper lip of the white felty calyx soon inflated to a sphere (Mediterranean, Canaries).

Woolly Clover
Trifolium tomentosum

Narrow-leaved Clover

Trifolium angustifolium · Leguminosae | Apr–Jul

Clover-species often have large stipules, which form a membranous sheath around the leaf stalk and end in a free, in this case pointed, section.

Identification 0.1–0.6(–0.9)m. Annual, erect, downy Clover. Leaves trifoliate, with long, narrowly linear leaflets, in this species unusually with entire margin, stipules hairy, spreading. Flower heads stalked, narrowly egg-shaped, 2–13cm long. Corollas pink, 7–15mm, the calyx scarcely projecting. Calyx teeth roughly equal, awl-shaped, with long cilia, spreading at fruiting time.
Habitat Cultivated land, waste ground, roadsides, garigue.
Other species The Purple Clover is similar, but in the long heads the bright purplish-red, 16–25mm long corollas far exceed the calyx. The lower calyx teeth twice as long as the upper (S France, Balkan Peninsula to Egypt).

Purple Clover
Trifolium purpureum

At fruiting time the calyx teeth with reddish-brown tinged inner side stick out in the shape of a star.

The species is easily recognised by its narrow grass-like leaves.

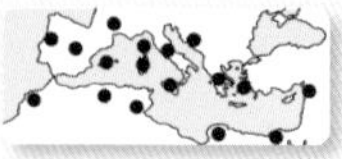

Italian Sainfoin, French honeysuckle

Hedysarum coronarium · Leguminosae | Mar–Jun

The Italian Sainfoin is often grown over large areas as a fodder plant. It is actually only native in the western Mediterranean, but also widely established in the wild and naturalised.

Sulla spinosissima

Identification 0.3–1m. Perennial, ± erect, herbaceous plant. Leaves with 5–11 broad egg-shaped leaflets, almost hairless above, with flattened hairs beneath. The 12–20mm long, crimson flowers in long-stalked, dense, oblong racemes. Pods of 2–4 flat, rounded segments, sometimes with small prickles.
Habitat Cultivated land, waste ground, roadsides.
Other species Formerly included in *Hedysarum*, the annual, prostrate *Sulla spinosissima* has leaves with 9–17 narrow leaflets. Flowers pink, 8–11mm long, in heads of 1–8. Pod segments hairy and with hooked, yellowish or reddish-brown prickles (Mediterranean).

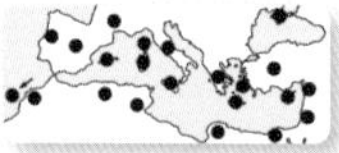

Asparagus Pea, Winged Pea

Lotus tetragonolobus · Leguminosae | Mar–Jun

The plant was formerly widely cultivated as green fodder for cattle and also as a vegetable, because the young fruits of the Asparagus Pea are edible when prepared like asparagus or peas.

L. biflorus

Identification 0.1–0.6m. Annual, spreading, softly hairy plant. Leaves 5-parted, the terminal leaflets broad reverse egg-shaped to rhomboid, the 2 lower ones smaller, pointed egg-shaped, like stipules. Flowers 1–2, roughly as long-stalked as the accompanying leaf. Corolla scarlet, 15–22mm, wings often with a dark spot. Calyx 2-lipped, the teeth twice as long as the tube.
Habitat Cultivated land, roadsides, grassy areas.
Other species The prostrate *L. biflorus* has 1–2 yellow to orange flowers, their stems twice as long as the accompanying leaf. Pods hairy, with 4 roughly 1mm wide wings (central Mediterranean).

This species with its conspicuous crimson flowers rarely forms fruits.

The hairless pods have four 2–4mm wide, often wavy wings.

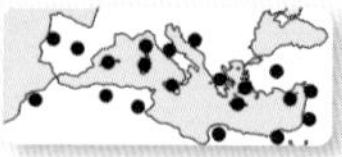

Mediterranean Thyme

Thymbra capitata · Labiatae | May–Oct

Mediterranean Thyme is traditionally used like Common Thyme, but the taste is stronger as the active ingredient, the essential oil, has a different composition and is considered inferior.

Flower head of Mediterranean Thyme

Identification 0.2–0.5m. Strongly aromatic, dwarf shrub often growing in spherical bushes, with white felty, rounded branches. Leaves slightly boat-shaped, not inrolled at the edge, ciliate at the base, otherwise hairless, often shed during the dry season (winter leaves). In their axils bunches of smaller leaves which survive the dry season (summer leaves). Flower heads with greenish, ciliate bracts, arranged like roof tiles, covering the calyxes. Corolla pinkish-violet, up to 10mm long. Calyxes flat on the back, unlike the Common Thyme.
Habitat Garigue, rocky ground, mainly on limestone.

Garden Thyme, Common Thyme

Thymus vulgaris · Labiatae | Apr–Jul

Garden Thyme, Common Thyme is familiar as a culinary herb and medicinal plant. Preparations of Common Thyme are mainly used as an expectorant for coughs; the essential oil, due to its high thymol content with strong germicidal properties, is also used externally.

Thymus striatus

Identification 0.1–0.3m. Strongly aromatic dwarf shrub with erect, brown, 4-angled branches. The grey-green leaves are short-stalked, linear to elliptical, with inrolled margin, white felty beneath, not ciliate. Flowers with 4–6mm long, pale violet corolla, in false whorls of 3–6 with leaf-like bracts.
Habitat Garigue on limestone, widely cultivated.
Other species *Thymus striatus* has creeping shoots with linear leaves with glandular dots, not inrolled at the edge, spreading bristly at the base. Flowers whitish to pink (Italy, Balkan Peninsula).

Mediterranean Thyme often occurs in stands.

The flowers occur in false whorls arranged in heads or interrupted spikes.

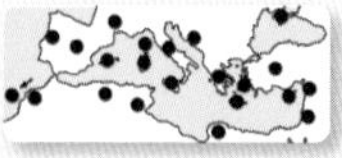

Common Snapdragon

Antirrhinum majus · Plantaginaceae | May–Nov

Red- and yellow-flowering Snapdragon species are popular ornamental plants in a wide range of varieties which have established in the wild as far as the warm parts of non-Mediterranean Europe.

Identification 1.5–2m. Biennial to perennial, sturdy, woody plant with opposite, alternate above, simple, ovate-lanceolate, leaves, wedge-shaped at base. Purplish-red flowers in terminal racemes, corolla 25–45mm long, with 2-lipped upper and 3-lipped lower lips.
Habitat Rocky slopes, walls. Only native in western Mediterranean.
Other species The Sicilian Snapdragon has a 17 to 25mm long, pale yellow corolla with dark yellow throat, the upper lip sometimes with violet markings. Leaves linear to narrowly elliptical (0.2–0.6m, Dec–Jun, Sicily, Malta, established in the wild elsewhere).

Sicilian Snapdragon
A. siculum

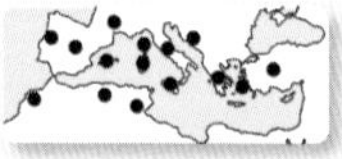

Evergreen Honeysuckle, Mediterranean Honeysuckle

Lonicera implexa · Caprifoliaceae | Apr–Jul

It is difficult to assign Honeysuckles to a flower colour. They are often tinged with red outside, inside often white when first open, then the colour changes to yellow.

Identification 1–3m. Evergreen subshrub with twining, reddish shoots. Leaves leathery, hairless, smooth at the edge, slightly transparent, short-stalked and opposite, the upper broadly joined at base. Strongly scented, stalkless flowers 2–6(–9). Corolla 2.5–4.5cm long, with narrow tube and 2-lipped limb. Stamens scarcely projecting. Berries red, probably slightly poisonous.
Habitat Evergreen thickets, woods, ornamental shrub.
Other species The deciduous Etruscan Honeysuckle has ± short-stalked, hairy leaves, at least the topmost pair joined. Inflorescence stalked, with 8–12 white flowers, stamens projecting (Mediterranean).

Etruscan Honeysuckle
Lonicera etrusca

The throat is closed by two white or yellow appendages on the lower lip.

The narrow corolla tube has two lips, with a 4-lobed upper and a simple lower lip.

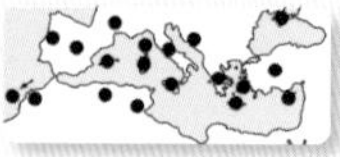

Annual Valerian
Centranthus calcitrapae

Red Valerian

Centranthus ruber · Caprifoliaceae | Apr–Oct

The Red Valerian was once considered as a medicinal plant: its roots contain relaxing valepotriates. However, the compounds were taken off the market due to possible damage to the genetic make-up.

Identification 0.3–0.8m. Perennial, hairless, ± blue-green plant. Leaves opposite, pointed ovate-lanceolate, mostly entire. Flowers in cymes, the pink corolla with 7–10mm long, narrow tube and unevenly 5-lobed limb, the thin spur 4–7mm. A single, projecting stamen.
Habitat Rock crevices, scree, walls.
Other species In contrast the Annual Valerian is a delicate plant. Lower leaves usually simple, rounded, toothed, upper pinnately parted with linear segments. Flowers pink or white, corolla tube 1–3mm, with sack-like protrusion at base (Mediterranean, Canaries).

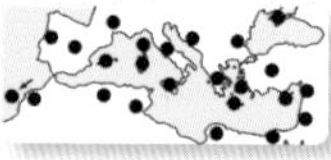

Wild Gladiolus
Gladiolus illyricus

Field Gladiolus

Gladiolus italicus · Iridaceae | Mar–Jun

Gladioli are familiar as large ornamental plants. The species in the Mediterranean, on the other hand, are delicate and sometimes occur in large stands.

Identification 0.4–1.2m. The perennial, tuberous plant has narrow lance-shaped leaves, up to 16mm across. Flowers 6–16 in loose, one-sided, simple spikes, supported by 1 long and 1 short floral bract. Perianth pink, the 6 lower segments fused into a short tube, the uppermost one distinctly separate. Anthers longer than the filaments.
Habitat Farmland, especially cornfields.
Other species The leaves of the Wild Gladiolus are 4–10mm wide, the 3–10 flowered spike sometimes has a side branch. Uppermost perianth segment only slightly longer than the lateral ones. Anthers as long as the filaments or shorter (0.2–0.6m, garigue, woods, Mediterranean).

The attractive Red Valerian is also known as an ornamental plant elsewhere in Europe.

The uppermost perianth segment is broader and longer than the side ones.

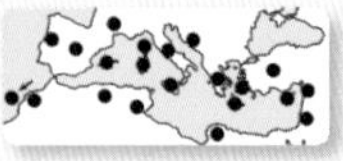

Dense-flowered Orchid

Neotinea maculata · Orchidaceae | Mar–May

The division of Orchidaceae according to genera and species is still in a state of flux due to recent research. Nowadays 4 species are included in *Neotinea*, of which 3 were previously considered to be species of *Orchis*.

Identification 0.1–0.3(–0.4)m. Orchid with 2–3 basal, often blackish-brown spotted, oblong or egg-shaped leaves. Flowers with scent of vanilla, in many-flowered spikes. 5 sepals incurved to form a hood, dirty pink or greenish white, 3–4mm long. Lip often mottled, 3-lobed, the middle lobe longer, 2–3 toothed, angled obliquely downwards. Spur up to 1.5mm, blunt.
Habitat Evergreen woods, up to the montane zone.
Other species The Burnt-tip Orchid has unspotted leaves. Flowers honey-scented, with hood dark red outside, middle lobes of the tripartite lip bifid, pointing forwards (S Europe, northwards to Sweden).

Burnt-tip Orchid
Neotinea ustulata

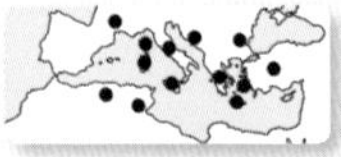

Milky Orchid

Neotinea lactea · Orchidaceae | Feb–Apr

Even if this orchid owes its name to the white background colour of its flowers, the red mottling of the lip is the dominant impression.

Identification 0.1–0.2m. Orchid with 4–6 basal, unspotted leaves, 2–3 leaves clasping the sturdy stem like a sheath. Inflorescence egg-shaped, dense. 5 white to greenish sepals incurved to form a hood, the outer with short tips turned outwards. Lip 6–12mm, deeply 3-lobed, ± toothed, with sinuate middle section. Spur thick, 5–7mm long, pointing downwards.
Habitat Grassy areas, maquis, open woods.
Other species The Three-toothed Orchid has a dense, egg-shaped inflorescence. Outer sepals tapering gradually to a point, lip with abundant red mottling, the middle lobe with distinct denticles (0.15–0.45m, Mar–Jun, Mediterranean, rarely to Central Europe).

Three-toothed Orchid
N. tridentata

The Dense-flowered Orchid is the most inconspicuous of the Mediterranean orchids.

The robust, broadly ovate-lanceolate leaves are characteristic of this species.

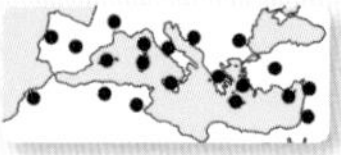

Pink Butterfly Orchid

Anacamptis papilionacea · Orchidaceae | Feb–May

A number of orchids are now assigned to the genus *Anacamptis*, which previously only had one species, *Anacamptis pyramidalis*. A characteristic which they share is the undivided middle lobe of the lip.

Identification 0.2–0.4m. Basal leaves narrow lance-shaped, channelled. Inflorescence egg-shaped, ± loosely 3–15-flowered. Sepals brownish-purple, with dark veins, roughly forming a hood. Lip 11–18mm, spread out like a fan, only indistinctly tripartite, irregularly denticulate at the edge, white to pink. Spur 12–14mm, bent downwards.
Habitat Grassy areas, maquis, open woods.
Other species The inflorescence of the Pyramidal Orchid is cone-shaped to egg-shaped. Flowers purplish-red, pink or white, lateral sepals spreading, the others forming a hood. Lip 6–10mm long, flat 3-lobed, with 2 spurs at the base (Mediterranean, rarely to Central Europe).

Pyramidal Orchid
Anacamptis pyramidalis

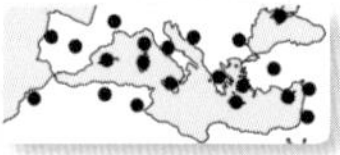

Green-winged Orchid

Anacamptis morio · Orchidaceae | Mar–May

This now includes 6 subspecies which were previously classified under *Orchis*.

Identification 0.1–0.3(–0.5)m. Inflorescence 3–20-flowered, 5 perianth segments in the form of a hood, free at the apex, pink to dark violet, with green veins inside. Lip broader than long, usually with dark spots, 3-lobed. The robust ssp. *morio* with a thick, cylindrical, rounded spur is the only subspecies which occurs from S Europe to N and W Europe. The ssp. *picta* has a more slender inflorescence, the side lobes longer and narrower than the middle lobe, spur more delicate (SW Europe). The ssp. *caucasica* has smaller side lobes (Balkan Peninsula to Turkey). In the ssp. *longicornu* the side lobes of the lip are darker, the middle lobe with white, spotted band, spur broadened (western Mediterranean).
Habitat Grassy areas, garigue, open woods.

Anacamptis morio ssp. *longicornu*

The fan-shaped lip is marked with dots and lines.

The photos show ssp. *picta* on the left and ssp. *caucasica* on the right.

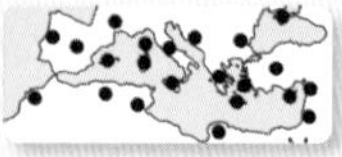

Bug Orchid

Anacamptis coriophora · Orchidaceae | Apr–Jun

The name derives from the "bug stink" of the ssp. *coriophora*, which also occurs infrequently as far north as Central Europe. The dominant ssp. *fragrans* of the Mediterranean, on the other hand, has a pleasant smell like vanilla.

Holy Orchid
Anacamptis sancta

Identification 0.2–0.5m. Basal leaves 2–4, tapering to a lance-shaped point and folded. Inflorescence oblong, dense-flowered, with 1-veined, long bracts. Flowers of the ssp. *fragrans* shown are brownish-red and green. The spotted lip 5–10mm long, 3-lobed, the middle lobe distinctly longer than the side ones, middle lobe in ssp. *coriophora* scarcely longer.
Habitat Grassy areas, maquis, open woods.
Other species The Holy Orchid has long pointed sepals and 3–5-veined lower bracts. Lip without spots, the side lobes coarsely toothed. Spur bent downwards (Crete, Aegean to Israel).

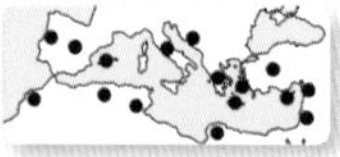

Naked Man Orchid

Orchis italica · Orchidaceae | Feb–May

The Naked Man Orchid was first described from Italy. It is common throughout the Mediterranean, often forming large stands.

Monkey Orchid
Orchis simia

Identification 0.2–0.5m. Orchid with broad lance-shaped spotted or unspotted leaves, often with wavy edges. Inflorescence egg-shaped, compact and dense-flowered, opening from the bottom upwards. Sepals pale pink, tapering to a point, incurved. Lip 14–17mm, white or pink, deeply divided into three, the middle lobe subdivided, with elongated denticles. Spur 6mm, pointing downwards.
Habitat Grassy areas, maquis, open woods.
Other species The Monkey Orchid has smooth, unspotted leaves. Flowers opening from above downwards, "monkey-like", all sections of the lip blunt and bent, dark red. Spur somewhat thickened (Mediterranean).

The pointed sepals form a beaked hood.

All segments of the lip, which is spotted and lighter in the middle, are linear and pointed.

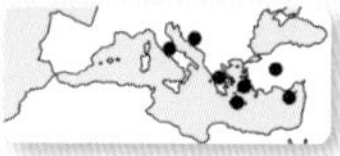

Four-spotted Orchid

Orchis quadripunctata · Orchidaceae | Mar–Jun

The name of the Four-spotted Orchid is justified, even if often only 2–3 dark red points are visible, 2 others are hidden in the entrance to the spur.

Identification 0.1–0.3m. Basal leaves oblong-lanceolate, mostly spotted. Flowers white, pink to purplish-violet, in a loose head of 8–35 on an often dark red-tinged stem. Outer sepals rounded and spreading, the 2 lateral inner ones smaller, incurved. Lip 6–8mm, roughly uniformly 3-lobed, flat, paler at base. Spur long and thin, 10–12mm, ± pointing downwards.
Habitat Rocky ground, grassy areas, garigue, on limestone.
Other species In the Anatolian Orchid the 2 outer sepals stick out, the middle one bends over the 2 inner ones. Lip 3-lobed, with a pale middle band with 2 rows of spots. Spur ± pointing upwards, 15–25mm (woods, thickets, from Crete and the Aegean eastwards).

Anatolian Orchid
Orchis anatolica

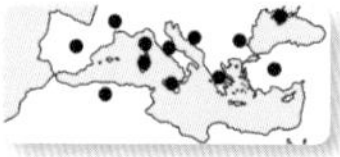

Lady Orchid

Orchis purpurea · Orchidaceae | Apr–Jun

The brownish-red dots on the lip are a conspicuous feature of this species: on closer inspection they are seen to comprise small bunches of hairs.

Identification 0.3–0.8m. Imposing Orchid with large, broad lance-shaped, unspotted basal leaves. Inflorescence profuse and dense, sepals brownish-purple veined outside, tightly hooded. Lip 15–20mm long, white or pink, deeply 3-lobed, the middle lobe broader than the often narrow lateral ones, subdivided and with a denticle in the middle. Spur pointing downwards, 4–7mm.
Habitat Thickets, lawns.
Other species The Man Orchid, previously known as *Aceras anthropophorum*, has a long inflorescence with yellowish to reddish-brown flowers without spurs. Sepals with a dark margin, hooded, lip with 2 narrow lateral lobes and longer, deeply divided middle lobe (Mediterranean).

Man Orchid
Orchis anthropophora

The lip is flat, broader than long and roughly equally 3-parted

The darker coloured margin of the lip is finely lobed.

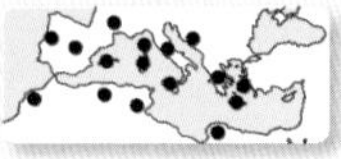

Tongue Orchid

Serapias lingua · Orchidaceae | Mar–May

A key characteristic of *Serapias* species is the divided lip of the flower. It consists of a rear section with erect side lobes, partially hidden in the greyish-violet hood of the sepals, and the front, tongue-shaped section.

Identification 0.1–0.4m. Stem with linear-lanceolate leaves. Flower spikes loosely 2–8-flowered, bracts greyish-violet, roughly as long as the flowers. Front part of the lip usually reddish, pointed, angled downwards or obliquely forwards from the slightly narrowed base, at most weakly hairy, 10–18mm long and 5–10mm wide.
Habitat Grassy areas, maquis, woods.
Other species The Small-flowered Tongue Orchid has few small flowers. Front lip only 5–10mm long and 3–5mm wide, usually sharply recurved. 2 dark parallel callus ridges at base (Mediterranean, Canaries).

Small-flowered Tongue Orchid
S. parviflora

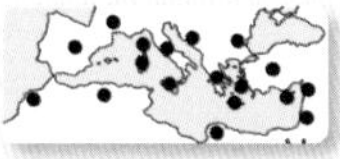

Long-lipped Tongue Orchid, Ploughshare Orchid

Serapias vomeracea · Orchidaceae | Mar–Jun

The 12–27 (depending on species definition) polymorphic Tongue Orchids are largely confined to the Mediterranean region.

Identification 0.15–0.6m. A large Tongue Orchid with folded and keeled leaves. The silvery-grey, dark purple-veined bracts distinctly longer than the 3–12 large flowers. Front lip pointing downwards to recurved, yellow-brown to rusty-red, 17–35mm long and 9–14mm wide, with dense long hairs at base. 2 parallel callus ridges at the base.
Habitat Garigue, grassy areas, plantations.
Other species The Heart-flowered Tongue Orchid also has large flowers. Bracts somewhat shorter than the flowers. Front lip heart-shaped, dark red to blackish-purple, 20 to 32mm long and 16–23mm wide, densely hairy roughly to the middle. 2 spreading callus ridges at the base (Mediterranean).

Heart-flowered Tongue Orchid
S. cordigera

The usually undivided dark callus ridge of this species is visible from outside.

Very few species of Tongue Orchids have such large attractive flowers.

Giant Orchid

Himantoglossum robertianum · Orchidaceae | Jan–May

This species was previously classified in a separate genus, *Barlia*. Based on molecular genetic studies it is now placed with the Lizard Orchids.

Identification 0.25–0.8m. Leaves egg-shaped to oblong, at the base of a strong stalk. The greenish to reddish or brownish, red spotted flowers in long, dense spikes. 5 perianth segments incurved to form a loose hood. Lip up to 2cm long, wavy at the edge, 3-lobed with bifid middle lobe.
Habitat Grassy areas, maquis, open woods.
Other species The Lizard Orchid is also large. Inflorescence many-flowered, with strong goaty smell, lip 3.5 to 6.5cm long, "strap-shaped", divided up to 7mm at the apex, greenish to brownish, spirally inrolled, later spirally twisted (May–Jul, western Mediterranean).

Lizard Orchid
H. hircinum

Yellow Bee-orchid

Ophrys lutea · Orchidaceae | Feb–May

The perennial tuberous *Ophrys* species also have their centre of distribution in the Mediterranean, only a few extend to more northern parts of Europe.

Identification 0.1–0.3m. Basal leaves 3–6, lance-shaped. Stem with 2–10 flowers. Outer sepals (yellow)green, the upper bent forwards, the lateral spreading. Lip rounded to oblong, up to 16mm, 3-lobed, the middle lobe often retuse, brown in the middle with grey-blue speculum, the yellow margin in ssp. *lutea* shown here 6mm across, narrower in the other subspecies.
Habitat Grassy areas, garigue, open woods.
Other species The Sombre Bee Orchid is also very variable. Lip oblong, 13–23mm, 3-lobed, velvety, dark reddish-brown, with narrow, hairless, yellow or brown limb. The blue-grey or blue-violet speculum divided in 2 (Mediterranean).

Sombre Bee Orchid,
Brown Bee Orchid
Ophrys fusca

The markings on the lip of the Giant Orchid are very variable.

The width of the yellow margin of the lip varies depending on the subspecies.

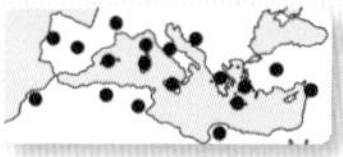

Mirror Orchid

Ophrys speculum · Orchidaceae | Feb–May

The polymorphic genus *Ophrys* has around 30 species with numerous subspecies or up to 250 species, depending on your viewpoint.

Identification 0.1–0.25m. Stem with 2–8 flowers. Outer sepals green, often striped brownish-violet. Lip 12–16mm, scarcely curved, 3-lobed with rounded middle lobe, the margin densely brown hairy and a hairless, shiny blue speculum with yellow margin in the middle. In ssp. *regis-ferdinandii* (Eastern Aegean) the lobes narrower, strongly curved.
Habitat Grassy areas, garigue, open woods.
Other species The Bumblebee Orchid has the smallest, only up to 10mm long lip. Side lobes with shaggily hairy tubercles, middle lobe strongly curved, the appendage bent backwards. The inconspicuous, shield-shaped or 2-parted speculum bluish to brownish (Mediterranean, Canaries).

Bumblebee Orchid
Ophrys bombyliflora

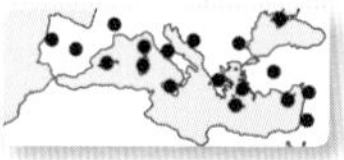

Early Spider Orchid

Ophrys sphegodes · Orchidaceae | Feb–Jun

The species name, which comes from Greek, actually means wasp-like. However, as there was no noticeable similarity to wasps, the species was later called Spider Orchid.

Identification 0.1–0.5m. Stem with 3–12 flowers. Outer perianth segments yellowish green or pink. Lip 8–16mm, rounded to egg-shaped, mostly entire, ± tubercles, hairy on outer margin, hairless inside, light to dark brown, with H-shaped, bluish marking, often with light margin. Polymorphic, with numerous subspecies.
Habitat Grassy areas, garigue, open woods.
Other species In Reinhold's Bee Orchid the sepals are usually pinkish-violet. Lip blackish-purple, 3-lobed, side lobes densely hairy, middle lobe oval, with appendage at front. The speculum composed of 2 separate or joined, white or violet patches with white margin (Balkan Peninsula to Middle East).

Reinhold's Bee Orchid
Ophrys reinholdii

Ssp. *speculum* shown on left, ssp. *regis-ferdinandii* on right.

Ssp. *sphegodes* shown on left, ssp. *mammosa* on right.

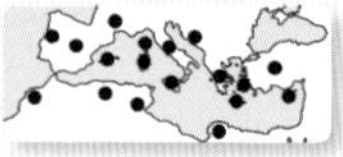

Sawfly Orchid

Ophrys tenthredinifera · Orchidaceae | Mar–Jul

The insect-like form of the flower and a special scent compound attract males of certain species of *Hymenoptera* to an attempted mating. This results in pollen packets, usually on the insect's head, being transferred to the next flower.

Identification 0.1–0.3m. Basal leaves 3–5, broad lance-shaped. Stem with 3–8 flowers. Outer sepals wide, usually pink to red, spreading, concave. Lip 11–16mm long, with 2 weak tubercles at base, reddish-brown in the middle, with hairy, yellowish margin. The speculum small, shield-shaped, grey-violet with pale margin.
Habitat Grassy areas, garigue, open woods.
Other species In the Late Spider Orchid the outer sepals often greenish. Lip trapezium-shaped, dark brown in the middle, with long hairs at the edges, with clearly divided speculum (Mediterranean).

Late Spider Orchid
Ophrys fuciflora

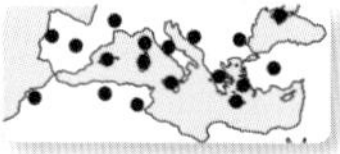

Woodcock Orchid

Ophrys scolopax · Orchidaceae | Mar–Jun

A characteristic of the Woodcock Orchid is the broad, upwards-pointing appendage on the labellum.

Identification 0.1–0.4m. Basal leaves 3–7, broad lance-shaped. Stem with 2–12 flowers. Outer perianth segments whitish to pink, mostly reflexed. Lip brown, oval, 6–13mm, deeply 3-lobed, tubercles of densely hairy side lobes in ssp. *scolopax* up to 4mm long. The speculum brownish-violet with pale margin. The ssp. *cornuta* with 6–12mm long tubercles (eastern Mediterranean).
Habitat Grassy areas, garigue, open woods.
Other species In the Bertoloni's Bee Orchid the oblong-ovate lip is 12–18mm long, mostly entire, blackish-purple and densely hairy. The front part saddle-shaped and bent upwards, notched at front and with erect, yellowish appendage, the bright blue speculum shield-shaped (central Mediterranean).

Bertoloni's Bee Orchid
Ophrys bertolonii

The lip has a small, hairless, upward-pointing appendix.

Of the 6 subspecies, ssp. *scolopax* is shown on the left and ssp. *cornuta* on the right.

Love-in-a-Mist

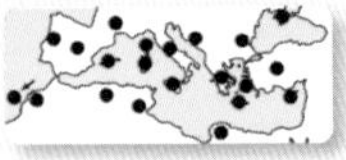

Nigella damascena · Ranunculaceae | May–Jul

The aromatic Black Cumin seeds are familiar as a spice used in cooking. These are obtained from cultivated Black Cumin *Nigella sativa*.

Black Cumin
Nigella sativa

Identification 0.1–0.5m. Leaves of the annual plant 2–3 times pinnately parted with narrow, pointed segments. Flowers surrounded and exceeded by finely divided floral bracts. Perianth segments egg-shaped, light blue to white, 12–20mm long, with short claw and 2-parted nectaries at base. Carpels forming a spherically swollen, hairless fruit. Style ± horizontally spreading.
Habitat Cultivated land, waste ground, also ornamental plant.
Other species The Black Cumin has no floral bracts. Flowers with 5 whitish to bluish sepals. Carpels fused up to the apex, rough warty. Style erect (established in the wild and naturalised in eastern Mediterranean, native W Asia).

Blue Flax

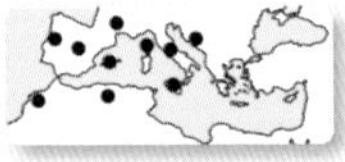

Linum narbonense · Linaceae | Jun–Jul

Several blue-flowered species of flax occur in the Mediterranean, most have smaller flowers, such as the widely distributed, delicate biennial Pale Flax *Linum bienne*.

Hairy Pink Flax
Linum pubescens

Identification 0.3–0.5m. Perennial, hairless flax, woody at base. Leaves narrow, linear to lance-shaped, tapering to a point. The bright blue flowers with 25–40mm long petals with darker veins at base, these 2.5–3-times as long as the pointed sepals. Capsule 5–6mm, spherical with fine apex.
Habitat Grassy areas, rocky ground.
Other species The Hairy Pink Flax is an annual with soft spreading hairs. Leaves broad lance-shaped. Petals pink, darker veined, blue at base, at 16–20mm twice as long as the sepals, these with simple hairs and with single glandular hairs at the edge (0.7–0.3m, Apr–Jun, garigue, waste ground, eastern Mediterranean).

Finely divided floral bracts extend beyond the perianth segments.

The sepals have a broad, membranous margin.

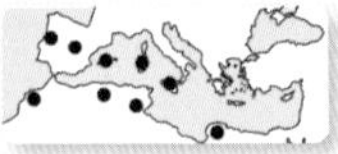

Garden Pimpernel, Flaxleaf Pimpernel

Anagallis monelli · Primulaceae | Mar–Jul

The Garden Pimpernel makes a wonderful show with its magnificent cushions of large, bright blue or vermilion, rarely white flowers.

Identification 0.1–0.5m. Creeping plant, woody at base. Leaves opposite stalkless, the upper also 3(–5), linear or egg-shaped. Flowers stalked, with flat spreading, 5-pointed, 12–25mm wide corolla. Filaments with purple or whitish hairs. Calyx 4–7mm, the 5 segments narrow, with membranous margin.
Habitat Waste ground, cultivated land, dunes.
Other species The Blue Pimpernel is an annual and always has blue flowers. Upper leaves lance-shaped, corolla only 5–8mm across, segments not overlapping, calyx in the opened flower visible from above over almost the whole length (Mediterranean, Canaries). The species is often confused with the blue-flowered variety of the Scarlet Pimpernel (*Anagallis arvensis* var. *caerulea*).

Blue Pimpernel
Anagallis foemina

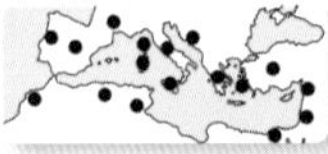

Sea Lavender

Limonium narbonense · Plumbaginaceae | Jul–Oct

Sea Lavenders occur throughout the entire region on flat and rocky coasts in over 300 species which are often difficult to distinguish. Many are restricted to very small sections of coastline and therefore endangered.

Identification 0.3–0.7m. Perennial plant with basal, lanceolate-spatula-shaped, long tapering leaves. The 5 petals bluish-violet, joined at base, 7–10mm long. Calyx 5-toothed, 5–6mm, with membranous, whitish to pale lilac limb.
Habitat Salt marshes of flat sea coasts.
Other species The Winged Sea Lavender or Statice has sinuately pinnatisect leaves. Stem with 4 wavy winged borders, which extend as pointed appendages at the nodes. Petals yellowish-white, calyx 12–14mm long, the membranous, bluish-violet limb almost entire (Apr–Aug, Mediterranean, Canaries, ornamental plant).

Winged Sea Lavender or Statice
Limonium sinuatum

The short corolla tube of both the blue and red flowers has a reddish ring.

The large inflorescence with loosely arranged branches has no wings.

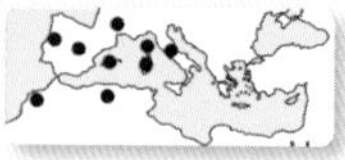

Intermediate Periwinkle

Vinca difformis · Apocynaceae | Feb–May

The Intermediate Periwinkle sometimes provides an impressive show of massed pale blue flowers. Its relatives have dark blue flowers, including the Lesser Periwinkle with flowers only 2cm across, which occurs in almost the whole of Europe. All species contain poisonous alkaloids.

Identification Perennial plant with up to 2m long, prostrate and up to 0.3m erect, flowering shoots. The evergreen leaves opposite, ovate-lanceolate, short-stalked. Flowers single, stalked in the upper leaf axils. Corolla pale blue, 3–4.5cm across, with narrow tube and flat limb. Calyx lobes hairless.
Habitat Damp shady scrub, edges of ditches.
Other species The Greater Periwinkle has distinctly ciliate leaf margins and calyx lobes. Flowers darker bluish-violet, 3–5cm across (western and central Mediterranean, Canaries, further naturalised).

Greater Periwinkle
Vinca major

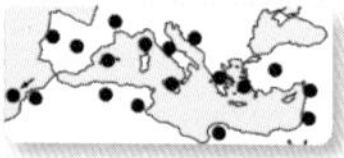

Tricolour Convolvulus

Convolvulus tricolor · Convolvulaceae | Mar–Jun

The flower colour of the Tricolour Convolvulus is unique. It is popular in cultivation and often runs wild. In some countries it is doubtful whether or not it is native.

Identification 0.2–0.6m. Annual or short-lived perennial, prostrate to erect hairy plant with stalkless, reverse egg-shaped to oblong leaves. Flowers single, their stems as long as the accompanying leaves or longer. Corolla funnel-shaped, 1.5–4cm long, 3-coloured. Sepals with long hairs, mostly distinctly 2-parted, with green segment above and colourless one below. Capsule hairy. 2 subspecies.
Habitat Cultivated land, waste ground, roadsides.
Other species The annual Small Blue Convolvulus is delicate. Leaves stalked, acute egg-shaped, heart to wedge-shaped at base. Flowers blue, with only 7–12mm long, distinctly 5-lobed corolla (Mediterranean, Canaries).

Small Blue Convolvulus
Convolvulus siculus

The 5 lobes of the corolla are asymmetrical and arranged obliquely like a propeller.

The corolla is yellow at the base, white in the middle and blue around the edge.

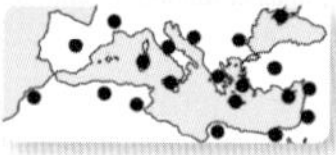

Dyer's Alkanet

Alkanna tinctoria · Boraginaceae | Apr–Jun

The root bark contains a mixture of red dyes called alkannin. This is still used for dyeing leather and cosmetics, but in Europe it is now prohibited as a food colouring.

Identification 0.1–0.3m. Prostrate to ascending, grey bristly hairy plant, woody at base. Lower leaves stalked, linear-lanceolate, upper stalkless with heart-shaped base. Flowers in initially dense, later elongated scorpioid cymes. The brilliant blue corolla hairless outside, with 5-lobed, 6–8mm wide, funnel-shaped limb.
Habitat Sandy and rocky coasts, waste ground.
Other species The Greek Alkanet is a stiffly hairy plant with glandless or glandular-sticky stems and calyxes. Leaves linear-oblong, with smooth margin, flat. Corolla yellow, 8–10mm across at the limb (rocky ground, open woods, up to the mountains, Greece, Bulgaria).

Greek Alkanet
Alkanna graeca

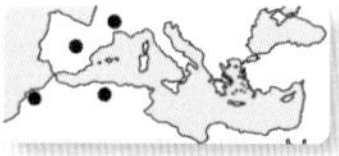

Shrubby Gromwell

Lithodora fruticosa · Boraginaceae | Mar–Jun

The leaves of most of the Boraginaceae are rough to the touch due to numerous bristly hairs which are either simple or sit on little tubercles. The flowers are often in scorpioid cymes.

Identification 0.2–1m. Much branched, small shrub. Leaves stalkless, linear-oblong, about 3mm wide, inrolled at the edge, with simple spreading bristly hairs and other spreading hairs on tubercles. Corolla blue, funnel-shaped, hairless in the throat, 10–18mm long, outside only sparsely bristly on the 5-lobed limb.
Habitat Garigue, open woods, mainly on limestone.
Other species *Lithodora hispidula* forms low cushions with its stiff branches. Leaves narrowly ovate-oblong, up to 4mm wide. Corolla about 12mm, hairless outside, white when opening, later pink to blue (Aegean, Crete, east to Syria, Cyprus, Libya).

Lithodora hispidula

The large cushions of Dyer's Alkanet on sandy beaches are endangered nowadays.

The flowers are magenta in bud but bright blue when open.

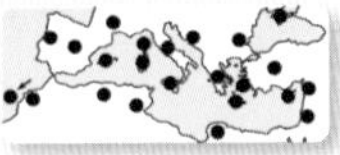

Italian Bugloss, Garden Anchusa

Anchusa azurea · Boraginaceae | Apr–Aug

The Italian Bugloss has been known as a garden plant since the 16th century. Native in the Mediterranean region, nowadays it is also naturalized in warm regions further north. It has also been known as *Anchusa italica*.

Identification 0.2–1.5m. Perennial, much branched plant with spreading, white bristly hairs, often on small tubercles. Leaves lance-shaped, the upper stalkless. Flowers with straight corolla tube, limb uniformly bluntly 5-lobed, flat, 10–15mm across, with a ring of white bearded coronal scales.
Habitat Agricultural land and waste ground, roadsides.
Other species The small, annual *Anchusella cretica* has ± curved toothed leaves. Flowers with slightly jointed corolla tube and unequal 5-lobed, 6–9mm wide, bright blue limb with narrow, violet margin (Italy, Sicily, Balkan Peninsula, Aegean).

Anchusella cretica

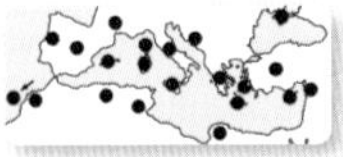

Borage

Borago officinalis · Boraginaceae | Apr–Sep

The fresh leaves with a cucumber-like smell and taste are used as a herb with salads but especially for pickling cucumbers (use in moderation on account of the alkaloid content). The flowers can be used as an edible decoration.

Identification 0.2–0.7m. Annual, bristly hairy plant. Lower leaves in a rosette, ovate-lanceolate, narrowed into the winged stalk. Flowers drooping, 2–3cm across, corolla sky-blue, more rarely white, with short tube and white coronal scales, the 5 pointed lobes spreading. Anthers blackish-violet, incurved.
Habitat Agricultural land and waste ground, originally only in the west, also cultivated and established in the wild.
Other species The white or bluish-violet flowers of *Symphytum creticum* have 5 narrow corolla lobes bent outwards (Mar–May, rock crevices, Greece, Crete).

Symphytum creticum

The blue flowers attract attention from far away.

You need to turn the downward-pointing flowers over in order to see them properly.

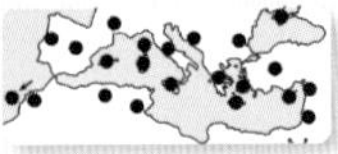

Blue Hound's-tongue

Cynoglossum creticum · Boraginaceae | Mar–Jul

The Blue Hound's-tongue is found throughout the Mediterranean region and is easy to identify.

Silvery-leaved Hound's-tongue, Yellowish Hound's-tongue *C. cheirifolium*

Identification 0.2–0.8m. Biennial, uniformly densely softly hairy plant. Leaves lance-shaped, lower ± in a rosette, narrowed into a stalk, the upper stalkless, half clasping the stem. Inflorescence without bracts, corolla with 5 rounded lobes, 5–7(–9)mm across. Fruits inclined downwards, fruitlets 5–7mm, curved, without thickened margin, covered with dense barbed bristles.
Habitat Waste ground, open, grassy areas.
Other species The Silvery-leaved Hound's-tongue has grey-white felty hairs. Flowers pale red or cream, with bracts. Fruits erect, fruitlets with thickened margin, with barbed bristles or more rarely almost smooth (western and central Mediterranean).

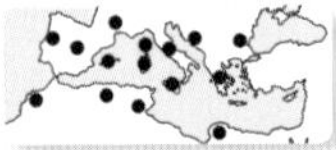

Apple of Sodom, Devil's Apple

Solanum linnaeanum · Solanaceae | May–Oct

Both the English and the older scientific name *Solanum sodomaeum* refer to the unattractive berries and the city of Sodom which, according the Bible, was destroyed because of the sinful behaviour of its inhabitants.

Silvery-leaved Nightshade *Solanum elaeagnifolium*

Identification 0.5–1.5(–3)m. Extremely thorny, squarrosely branched shrub. Leaves stalked, pinnately parted to almost the mid rib, with rounded, wavy lobes. Flowers with bluish-violet, 5-lobed, flat, 2–3cm wide corolla. Berries whitish and marbled green at first, later dirty brown, 2–3cm. Poisonous.
Habitat Roadsides, ruderal areas, naturalised, native to S Africa.
Other species The Silvery-leaved Nightshade originates in South America. Up to 0.6m subshrub with single, reddish prickles (Mediterranean).

The much-veined corollas are pink at first, turning pale blue later.

Branches, leaf veins and even the sepals have yellow prickles.

Mandrake

Mandragora officinarum · Solanaceae | Sep–Apr

The fleshy taproot reaching deep into the soil and often divided into two was thought to resemble a human form. This "little mandrake man" was assigned magical powers, such as banishing evil spirits. The mandrake was known even in ancient times as a remedy for pain or to induce sleep. Poisonous.

Mandrake roots

Identification 0.1–0.2m. Leaves in a large rosette, elliptical to reverse egg-shaped. Flowers short-stalked, erect, with 5-pointed, 12–65mm long, blue to violet corolla. Calyx expanded at fruiting time, as long as the orange-yellow, egg-shaped or spherical, 5–40mm berries. The widespread Mediterranean plants have been called *Mandragora autumnalis* with the name *M. officinarum* restricted to plants with greenish-white, smaller flowers opening in spring (central and eastern Mediterranean).
Habitat Waste ground, roadsides, ruins.

Adriatic Bellflower

Campanula garganica · Campanulaceae | Apr–Sep

Around 250, often only locally distributed, bellflowers occur in the Mediterranean region, for which the species illustrated can only serve as examples of the wide range on the Balkan Peninsula or in Asia Minor.

Chimney Bellflower
C. pyramidalis

Identification 0.1–0.2m. Perennial, polymorphic and freely flowering plant, pendulous or prostrate to erect, hairless or with felty hairs, with non-flowering rosettes. Leaves stalked, egg-shaped to rounded kidney-shaped, crenate or serrate. Flowers with light to dark blue, 10–15mm long corolla.
Habitat Limestone rocks, old walls, also ornamental plant.
Other species The perennial Chimney Bellflower can be recognised by its erect, tall growth form (up to 1.5m). Corolla roughly 25mm, incised to the middle (rocks, walls, N Italy, Croatia to Albania).

The wrinkled, wavy-edged leaves that lie flat on the ground are conspicuous.

The flowers have a corolla two-thirds incised with spreading lobes.

Blue Anemone, Apennine Anemone

Anemone apennina · Ranunculaceae | May–Jun

In spring in the Mediterranean it is well worth visiting the oak, beech and spruce forests of the montane zone where, besides other spring flowers, the Blue Anemone is to be seen.

Broad-leaved Anemone
Anemone hortensis

Identification 0.1–0.25m. Perennial anemone with sparse flattened hairs. Basal leaves stalked, 3-parted with lobed segments, the leaves of the trifoliate whorls of floral bracts similar. Flowers with 9–16 blue to white sepals, downy at base beneath. The similar Balkan Anemone or Mountain Windflower *Anemone blanda* with hairless involucral bracts replaces this species in the east.
Habitat Woods of the montane zone, montane meadows.
Other species The Broad-leaved Anemone has a whorl of 3 undivided bracts. Perianth segments 12–19, pale or dark purple (Feb–Apr, cultivated land, garigue, central Mediterranean).

Blue Aphyllanthes, Blue Grass Lily

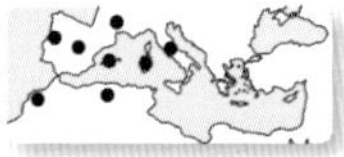

Aphyllanthes monspeliensis · Asparagaceae | Feb–Jul

The Blue Aphyllanthes is the only species in the genus *Aphyllanthes*.

Blue Aphyllanthes flowers

Identification 0.1–0.5m. Perennial, tussock-forming plant with rush-like, blue-green, ribbed and hairless, roughly 1mm wide stems, partly erect, partly arched over. Reduced leaves in the form of brownish, 3-veined, 3–8-cm long sheaths at the base. Flowers single or 2–3, terminal, with 6 blue sepals with a dark blue vein in the middle, 15–20mm long, spreading. Each flower with 1–5 blunt, calyx-like leaflets, joined below. 1–3 membranous, sometimes awned, 8–10mm floral bracts at their base.
Habitat Garigue, thickets, up to the mountains.

There are also lighter blue and white flowers amongst the bright blue.

At first glance this resembles a rush with blue flowers.

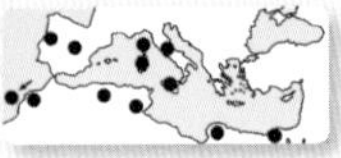

Portuguese Squill

Scilla peruviana · Asparagaceae | Mar–Jun

The Portuguese Squill is very attractive. The inflorescence is composed of 7–100 plus flowers, an indication of how variable the plant is. Despite the scientific name, it does not come from Peru.

Identification 0.2–0.5m. Perennial with up to 8cm large bulb. Leaves all basal, lance-shaped, 2–4(–6)cm wide, smooth at the edges, papillose or ciliate. Inflorescence a broad pyramid-shaped to hemispherical, dense raceme on a short, strong stalk, with awl-shaped, 2–8cm long bracts. The 6 perianth segments blue, 7–12mm long.
Habitat Damp pastures, thickets, open woods.
Other species The delicate Autumn Squill, by contrast, is inconspicuous, up to 0.2m high and flowers in autumn. Leaves 1–2mm wide, narrowly linear and channelled, only developing after the flowers. Flowers 6–25 in racemes without bracts, perianth segments bluish-violet, 4–5mm long, anthers dark violet (Mediterranean).

Autumn Squill
Prospero autumnale

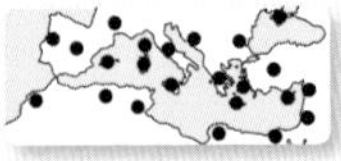

Common Grape Hyacinth

Muscari neglectum · Asparagaceae | Mar–May

Grape Hyacinths are typical early flowering plants and a number of species are popular for gardens. Only one delicate species flowers in autumn in the Mediterranean region, the Autumn Grape Hyacinth *Muscari parviflorum*.

Identification 0.1–0.35m. Perennial bulbous plant. Leaves linear to linear-lanceolate, all basal, 2–8mm wide, channelled. Flowers pitcher-shaped, in short, dense racemes on spreading or recurved, up to 5mm long stalks. Lower flowers fertile, 3.5–7.5mm long, blackish-blue with 6 white teeth, upper infertile, smaller and paler in colour.
Habitat Cultivated land, grassy areas.
Other species The Dark Grape Hyacinth has 5–15mm wide, flat to channelled leaves. The fertile flowers are blackish-violet and have teeth of the same colour (Italy, Sicily, Balkan Peninsula, Crete).

Dark Grape Hyacinth
Muscari commutatum

The flowers have blue, widened filaments with yellow anthers.

The fertile, dark blue flowers have white recurved teeth.

Requien's Sand Crocus
R. requienii

Sand Crocus

Romulea bulbocodium · Iridaceae | Jan–Apr

The Sand Crocus has a flower tube up to 1cm long and channelled, green leaves. Crocuses, on the other hand, have a longer flower tube, over 1.5cm, and mostly flat leaves with a white stripe.

Identification 0.05–0.25m. Variable bulbous plant with 3–5, channelled, up to 2mm wide leaves embracing half the stem. Perianth segments up to 4.5cm long, often tricoloured, yellow in the throat, whitish and violet above or flower completely violet. Stigmas overtopping the stamens. 2 wide membranous floral bracts below each flower.
Habitat Grassy areas, garigue, open thickets.
Other species Requien's Sand Crocus has 2, 4-angled, channelled leaves. Flowers dark violet, tube sometimes with white throat, segments ± blunt, up to 2cm long (Grassy areas and rocky ground, Corsica, Sardinia).

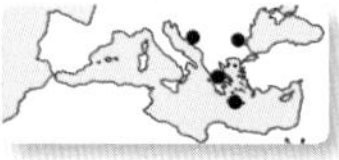

Silvery Crocus
Crocus biflorus

Sieber's Crocus

Crocus sieberi · Iridaceae | Mar–May

The genus *Crocus* occurs predominantly in the central and eastern Mediterranean region as numerous, often locally distributed species which are usually confined to mountainous regions.

Identification 0.05–0.1m. Bulbous plant with 1.5–6mm wide leaves with distinct, white stripe. Segments of the involucral bracts white, pale lilac or purplish-lilac, outside white or tinged purple, 1.5–4.5cm long, the throat yellow. Style yellow or orangey-red, about as long as the stamens.
Habitat Rocky slopes, grazed land, thickets.
Other species In the Silvery Crocus the 0.5 to 3.5mm wide leaves have only an indistinct white stripe. Perianth segments 1.5–4.5cm long, white, blue or lilac, the outer ones also with distinct dark veining, throat yellow. Very variable species (Balkan Peninsula, Crete).

The flowers of the Sand Crocus are very variable in colour.

While Sieber's Crocus flowers in spring, there are also numerous autumn-flowering species

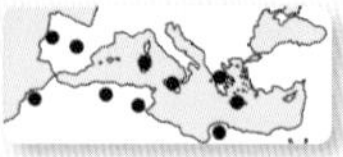

Flat-leaved Iris

Iris planifolia · Iridaceae | Nov–Apr

In Irises the perianth is fused into a tube at the base. The 3 outer segments are usually recurved and longer, the 3 inner often erect and shorter. 3 petal-like stylodia more or less cover the 3 stamens.

Identification 0.15–0.3m. Perennial plant, starts flowering in winter. Leaves 5–25mm wide, channelled, somewhat wavy at the edge. Flowers mostly 1–2, blue to violet, rarely white, with 6–12cm long tube, the 3 outer segments 4.5–7cm long, curved outwards from half way, the inner spreading, up to half the length.
Habitat Grassy areas, rocky ground, garigue.
Other species The variable Algerian Iris has 1–5mm wide, grooved leaves. Outer segments with violet, yellow and white veined lip, inner segments violet, erect (eastern Mediterranean, NW Africa).

Algerian Iris
Iris unguicularis

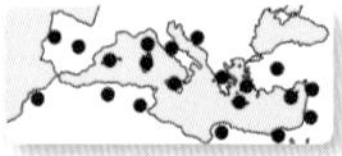

Barbary Nut

Moraea sisyrinchium · Iridaceae | Feb–May

The two Mediterranean species of *Moraea* were formerly classified in *Gynandriris* and are distinguished from *Iris* species – which have rhizomes or bulbs – by a 1.5–3cm wide, rounded, deep tuber with a fibrous coating, amongst other things.

Identification 0.05–0.5m. Perennial plant, the 1–3 channelled leaves with long sheath and 1.5–5mm wide, free lamina. Flowers 1–6, each in the axil of a 4–6cm long, dry membranous bract. Perianth light blue or violet, fused into a short tube below, the 3 outer segments recurved, about 3cm long, with a white patch, yellow at the base, the 3 inner erect.
Habitat Garigue, grassy areas, also on sand.
Other species *Moraea mediterranea* has only 1 narrow leaf (formerly called *Gynandriris monophylla*). Plant tiny, up to 6cm high, outer perianth segments 11–20mm (Greece, Crete, NE Africa).

Moraea mediterranea

The outer perianth segments have a yellow papillose strip in the middle.

Each flower of the Barbary Nut only opens for one afternoon.

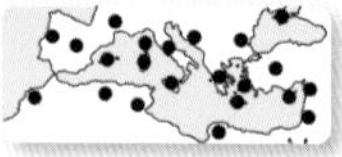

Sea Holly

Eryngium maritimum · Umbelliferae | Jun–Sep

Umbellifers usually have broad, umbrella-shaped inflorescences. It is rare for the umbels to be contracted into heads, as in the *Eryngium* species.

Identification 0.2–0.6m. Perennial, blue-green frosted plant. Basal leaves with 3–5-lobed, thorny-sinuately toothed lamina. Flowers ± stalkless, light blue, exceeded by 3-pointed spreading bracts, in 1.5–3cm large, initially spherical heads, surrounded by 4–7 broad, thorny toothed involucral bracts.
Habitat Sandy beaches, dunes.
Other species The Italian Eryngo has doubly pinnatisect basal leaves with thorny toothed, linear-lanceolate leaflets. Leafstalks of the stem leaves broadened, entire. Flower heads exceeded by 5–9 narrow, spiny involucral bracts (Rocky and grassy areas, eastern Mediterranean, westwards to Italy, Sicily).

Italian Eryngo
Eryngium amethystinum

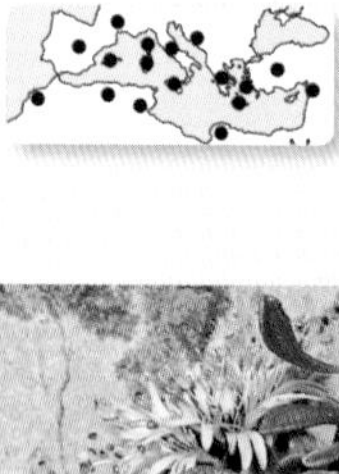

Shrubby Globularia

Globularia alypum · Globulariaceae | Oct–Apr

Globulariaceae also produce heads which are surrounded by bracts. The individual flowers are 2-lipped with 2-toothed upper lip and 3-parted lower lip.

Identification 0.2–1m. Erect branched, evergreen shrub, with only short-stalked, hairless and leathery, alternate leaves in clusters on the old branches up to the inflorescence. Flowers in this species with only very short upper lip, in spherical, 1–2.5cm wide, terminal and axial heads, surrounded by ciliate, broad egg-shaped, ± blunt bracts arranged like roof tiles.
Habitat Garigue, rocky ground, often forming stands.
Other species The woody Creeping Globularia spreads out over rocks. Leaves mostly in clusters, linear to spatulate, heads 6–15mm (0.03 to 0.06m high, May–Aug, mountain plant, Spain to NW Italy).

Creeping Globularia
Globularia repens

The anthers of the Sea Holly are also blue.

The reverse egg-shaped to oblong, pointed leaves often have three points.

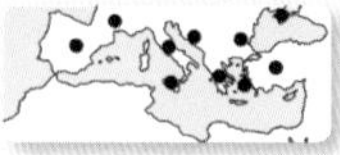

Globe Thistle

Echinops ritro · Compositae | Jul–Oct

Some *Echinops* are familiar as attractive ornamental plants. This variable species is also popular and sometimes established in the wild in Central Europe.

Identification 0.2–0.8m. Perennial, thistle-like plant with ± white felty stem. Leaves 1–2 times pinnatisect, segments with inrolled margin and 3–15mm long terminal thorn, upper side dark green, ± hairless, underside white felty. The spherical, blue, 2.5–4.5cm wide heads with 1-flowered florets, their involucral bracts in 3 rows, outer bristle-shaped, the remainder arranged like roof tiles.
Habitat Grazed land, rocky ground.
Other species The Spiny Globe thistle is much sturdier, up to 1.5m high. Leaves 2–3 times pinnatisect, glandular hairy above, white felty beneath and also glandular. Flower heads 3.5–7cm across, greyish-blue to greenish (eastern Mediterranean, westwards to Sicily, N Africa).

Spiny Globe Thistle
E. spinosissimus

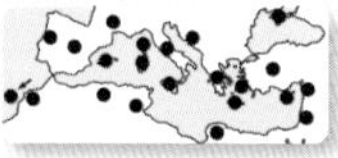

Cardoon

Cynara cardunculus · Compositae | Apr–Aug

The cultivated artichoke is familiar as a vegetable and medicinal plant. When making artichoke extracts, use is made primarily of the leaves of plants not yet in flower. They are used for disorders of the gall bladder and to lower cholesterol levels.

Identification 0.5–1.5m. Biennial plant with stiff, 1–2 times pinnatisect leaves, white woolly beneath, their segments with yellow thorns, sometimes in bunches. Flower heads 4–6cm across, with blue tubular florets. Fruitlets with dirty white, feathery pappus.
Habitat Pastures, roadsides.
Other species The Globe Artichoke has less thorny leaves. Flower heads 8–15cm across, involucral bracts blunt, with notched or thorny tip. The part eaten is the receptacle with the lower fleshy involucral bracts of the flower heads before they open.

Globe Artichoke
Cynara scolymus

From a botanical point of view the large heads consist of many individual florets.

The involucral bracts of the flower heads have long, yellow thorns.

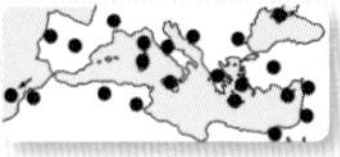

Narrow-leaved Lupin

Lupinus angustifolius · Leguminosae | Mar–Jul

The annual Lupins in the Mediterranean are delicate in contrast to the perennial species cultivated further north in Europe and elsewhere.

Identification 0.2–0.8m. Easily recognised as a Lupin with the fingered leaves, leaflets 5–9, linear-oblong, 2–5mm wide, often folded, only somewhat hairy beneath. Flowers all alternate in 10–20cm long raceme, with blue, more rarely pink or white, 12–16mm long corolla. Hairy, 4–7cm long fruits.
Habitat Agricultural land and waste ground, garigue, on acid, often sandy soils.
Other species The Hairy Lupin has 5–15mm wide leaflets, hairy on both sides. Blue flowers in 5–10cm long raceme, upper ± in whorls, lower alternate. Standard white in the middle, keel blackish-violet at the tip (Mediterranean, also cultivated).

Hairy Lupin, Bitter Blue Lupin
Lupinus micranthus

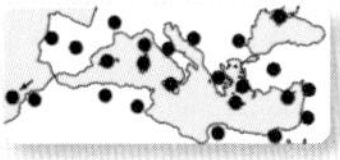

Pitch Trefoil

Bituminaria bituminosa · Leguminosae | Apr–Aug

The plant gets its name from the characteristic smell of bitumen, which is especially noticeable when the plant parts are crushed. The older name *Psoralea* refers to the former use for itchy skin conditions.

Identification 0.2–1.5m. Plant herbaceous, or woody at the base, ± with dense flattened hairs. Leaves long-stalked, trifoliate, the leaflets entire, with glandular dots, broad egg-shaped to narrow lance-shaped and tapering to a point. Flowers in 2–3cm broad heads, with stalks longer than the leaves, 7–30 bluish-violet, 15–20mm long flowers. Floral bracts 1–3-toothed, tapering to an awned point. Calyx with 5 unevenly long teeth. The single-seeded fruit remaining closed, with long bristly hairs and sword-shaped beak. Variable species.
Habitat Roadsides, ruderal sites, waste ground.

Fruit of the Pitch Trefoil

The lower lip of the calyx is about twice as long as the upper one.

The blueish-violet pea flowers often have whitish wings.

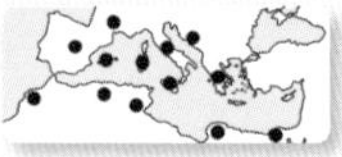

Coris

Coris monspeliensis · Primulaceae | Mar–Jul

The genus *Coris* is occasionally assigned to its own family, the Coridaceae.

Identification 0.1–0.3m. Herbaceous plant or with stem woody at base, with hairless or hairy, alternate, linear leaves, inrolled at the edge with thorny teeth or black glandular dots at the edge. Inflorescence densely spike-like, corolla deep pink or bluish-violet, 9–16mm long, unlike other Primulaceae with 5 unevenly long, 2-lobed tips. Calyx with double limb of 5 inner, wide triangular teeth, often with a blackish-purple dot, and 6–10(–14) unequal, thorny, recurved outer teeth. 3 subspecies.
Habitat Garigue, usually on limestone.

Coris flower buds

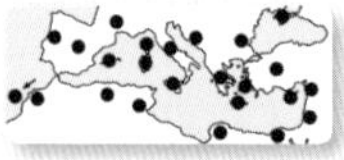

Purple Viper's Bugloss

Echium plantagineum · Boraginaceae | Feb–Jul

On first coming across the leaf rosette of this Viper's Bugloss by the roadside you are reminded of a Plantain (*Plantago*), to which the epithet *plantagineum* refers.

Identification 0.2–0.8m. Annual or biennial, soft bristly hairy plant. Basal rosette leaves stalked, egg-shaped, with distinct lateral veins. Upper stem leaves stalkless with ± heart-shaped base. Corolla magenta at first, later blue, (12–)20–30mm long, wide funnel-shaped with asymmetrical limb, almost 2-lipped, outside only slightly hairy at the edge and on the veins.
Habitat Roadsides, waste ground, ruderal areas.
Other species The Sand Viper's Bugloss is similar but basal leaves lance-shaped, without distinct lateral veins, stem leaves stalked or narrowed at base. Corolla 12–22mm long, outside ± uniformly hairy (on sand close to the coast, western and central Mediterranean).

Sand Viper's Bugloss
Echium sabulicola

Coris sometimes grows in large cushions.

Two of the five stamens project beyond the corolla.

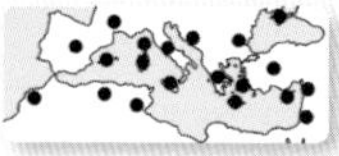

Chaste Tree, Monk's Pepper

Vitex agnus-castus · Labiatae | Jun–Nov

In ancient times and in the monasteries of the Middle Ages the peppery hot-tasting fruits were said to suppress sexual drive, thus giving the plant its name. Extracts of the plant are used nowadays in numerous proprietary medicines for regulating the female hormonal balance.

Fruits of the Chaste Tree

Identification 1–6m. Deciduous shrub, previously viewed as belonging to the Verbenaceae. Leaves opposite and decussate, long-stalked, fingered 5–7 times pinnate, the stalked leaflets lance-shaped, entire and white felty below, hairless above. Scented flowers in terminal, branched, spike-like inflorescences. Corolla 6–9mm long, 2-lipped, blue, more rarely pink or white, hairy outside. Small, fleshy, later reddish black fruits, enclosed by the felty hairy calyx.
Habitat River banks, damp sites, ornamental shrub.

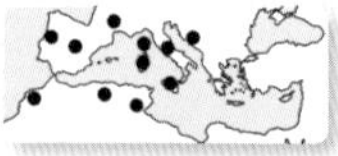

Shrubby Germander

Teucrium fruticans · Labiatae | Feb–Jun

A characteristic of the flowers of all *Teucrium* species is the lack of an upper lip. The lower lip 5-lobed with a long extended middle lobe.

Short-leaved Germander
Teucrium brevifolium

Identification 0.3–2.5m. Evergreen shrub with 4-angled, white felty branches. Leaves short-stalked, acute egg-shaped, flat, becoming hairless above, dark green, white felty beneath. Flowers stalked, in pairs in the upper leaf axils. Corolla pale blue to lilac, 15–25mm long, without a ring of hair in the throat, filaments far projecting. Calyx wide bell-shaped, white felty outside, hairless and green inside.
Habitat Coastal, evergreen thickets.
Other species The Short-leaved Germander has shorter, almost linear leaves, greyish-green on both sides with inrolled edges. Branches green or brown. Flowers with about 10mm long corolla, calyx sparsely hairy (0.3–0.6m, garigue, rock crevices, eastern Mediterranean).

The stamens project far beyond the two-lipped flower.

Shrubby Germander is popular as an ornamental shrub in the Mediterranean region.

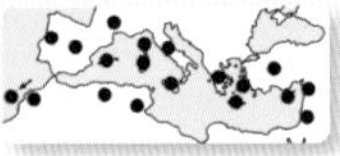

French Lavender

Lavandula stoechas · Labiatae | Jan–Jun

A tuft of oblong-ovate, mauve bracts projects beyond the inflorescence and serves as a device for attracting insects.

Identification 0.3–1.5m. Small, aromatic shrub with stalkless, oblong-lanceolate leaves, weakly grey felty on both sides with smooth margins. The dense inflorescence stalked or stalkless, with rounded, hairy bracts narrowing to a point, the upper enlarged to 20–40 mm, blueish-violet, without flowers, forming a tuft. Corolla 2-lipped, 6–8mm long.
Habitat Garigue, pinewoods, on siliceous rocks.
Other species The Fringed Lavender has a smaller tuft of 8–15mm long, mauve bracts. Flowers violet. Stem leaves lance-shaped, inrolled at the edge and regularly ± deeply crenate, green above, grey felty beneath (Spain, Balearics, NW Africa, abundant ornamental plant, established in the wild).

Fringed Lavender
Lavandula dentata

Common Lavender

Lavandula angustifolia · Labiatae | Jun–Aug

The essential oil from the flowers is frequently used in medicine, both internally and externally. With Spike Lavender (*Lavandula latifolia*) it forms a hybrid called Lavandin (*L. × intermedia*), the oil of which has the greatest share in the cosmetic industry.

Identification 0.2–1m. Low, strongly aromatic subshrub. Leaves linear-lanceolate, smooth at the edge and ± inrolled, white felty when young, later becoming hairless and green. Long stalked, spike-like inflorescences with pointed, broad egg-shaped, distinctly veined bracts. Corolla 10–12mm long, 2-lipped, bluish-violet, calyx about 5mm, greyishviolet.
Habitat Garigue, rocky ground, up to mountains, as ornamental, scent and medicinal plant in a range of varieties.
Other species The Cut-leaved or Fernleaf Lavender is only slightly aromatic. Leaves doubly pinnatisect, hairy, but green (Iberian Peninsula, S Italy, Sicily, N Africa).

Cut-leaved or Fernleaf Lavender
Lavandula multifida

Tiny, two-lipped, blackish-violet flowers are arranged on top of each other in rows.

Lavender is cultivated on a large scale, especially in Provence.

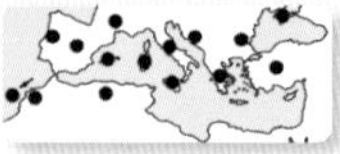

Common Sage

Salvia officinalis · Labiatae | May–Jul

The leaves of these two species of sage are used as a culinary herb and medicine. The essential oil is used in proprietary medicines, for example for inflammation of the mucous membranes of the mouth and throat.

Three-lobed Sage
Salvia fruticosa

Identification 0.2–0.6m. Subshrub with spreading grey-felty hairy branches. The stalked, thick leaves always simple, oblong-ovate, finely crenate at the edge, felty, becoming hairless above. Inflorescences loosely spike-like, corolla usually bluish-violet, 20–35mm long.
Habitat Garigue, rocky ground, Italy, Balkan Peninsula, further cultivated and naturalised.
Other species The Three-lobed Sage is a shrub with flattened grey felty branches. Leaves often with 2, more rarely up to 4 small, lateral auricles at base. Corolla 16–25mm long. The essential oil has a different composition (eastern Mediterranean, westwards to S Italy, Sicily).

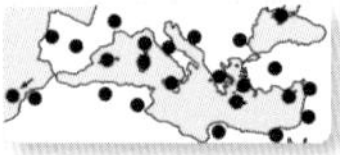

Wild Clary

Salvia verbenaca · Labiatae | Jan–Dec

There are a great many species of sage in the eastern Mediterranean: 90 species have been described from Turkey, whereas there are only about 25 in the whole of S Europe.

Red-topped Sage or
Annual Clary
Salvia viridis

Identification 0.1–0.8m. Perennial, herbaceous plant, usually with simple hairs, glandular hairs and stalkless glands in the upper part. The stalked leaves of the basal rosette ovate-oblong with a wrinkled surface, coarsely crenate and ± pinnatisect. Inflorescence spike-like, corolla 6–16mm, 2-lipped. Calyx with green or brownish-violet veins.
Habitat Waste ground, cultivated land, roadsides.
Other species The Red-topped Sage or Annual Clary is easy to recognise: there is often an elongated tuft of flowerless, green and violet bracts at the top of the flower shoot. Corolla with violet upper lip (Mar–Jun, garigue, waste ground, Mediterranean).

The two-lipped calyx is tinged brownish-violet with glandular dots.

The flowers of this variable species range from light blue to violet.

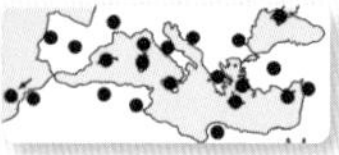

Rosemary

Rosmarinus officinalis · Labiatae | Jan–Dec

One of the highlights of the culinary herbs of the region, Rosemary leaves are used in many ways in cooking, but also serve as an appetite stimulating and digestive remedy.

Rosmarinus eriocalyx

Identification 0.3–1.5(–2)m. Evergreen, strongly aromatic shrub. Stalkless, narrow linear, leathery leaves with the edges inrolled beneath, 1.5–4cm long, green above, white felty beneath. Small, finely hairy inflorescences in the upper leaf axils. Corolla 10 to 12mm long, hairy outside, lower lip with large middle lobe, 2 curved blue filaments and style projecting a considerable way.
Habitat Garigue, maquis, also an ornamental plant.
Other species *Rosmarinus eriocalyx* is much rarer. Leaves shorter, 0.5–1.5cm, likewise green above. Inflorescence with long simple and branched hairs plus glandular hairs (S Spain, NW Africa).

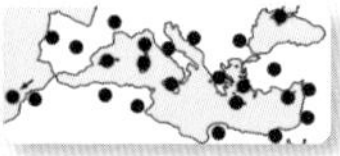

Branched Broomrape, Hemp Broomrape

Orobanche ramosa · Orobanchaceae | Feb–Sep

Broomrape species are obligate parasites without chlorophyll which live off the roots of other plants. Two examples are shown here of the approximately 75 Mediterranean species which are often difficult to distinguish.

Bean Broomrape
Orobanche crenata

Identification 0.05–0.3m. Mostly annual, glandular hairy Broomrape with branched stem and solitary scale leaves. Flowers in ± loose spikes, the lower sometimes stalked, each with an ovate-lanceolate subtending bract and 2 bracteoles attached to the 4-toothed calyx. Corolla 10–22mm long, light to dark blue.
Habitat Cultivated land, roadside, ruderal areas. On various host plants, especially hemp and tobacco.
Other species The Bean Broomrape has 1 narrow, pointed subtending bract for each flower and no bracteoles. Corolla whitish, tinged with mauve, with finely toothed limb, 25–30mm long (Mediterranean, Canaries).

Rosemary flowers are pale to deep blue, less often also white.

The corolla is 2-lipped, with a 2-toothed upper lip and 3-lobed lower lip.

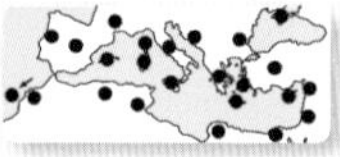

Maidenhair Fern

Adiantum capillus-veneris · Pteridaceae | Jun–Sep

Ferns spread via spores which develop in the sporangia, usually on the underside of the leaves. They are arranged in groups, known as sori.

Identification 0.1–0.6m. Perennial fern. Leaves with shiny, blackish-brown stem and 2–4 times pinnate lamina which is ovate-oblong in outline. Leaflets on hair-like stalks, bright green, tender, rounded, narrowed wedge-shaped, irregularly incised at end. Sporangia on the underside of the turned down edge lobes.
Habitat Shady damp limestone rocks, ornamental plant.
Other species The delicate, annual Jersey Fern dies off in April after the spores ripen. Leaves varied, lower rounded, successive 1–3 times pinnate (damp crevices in the ground, rocks, Mediterranean, Canaries).

Jersey Fern
Anogramma leptophylla

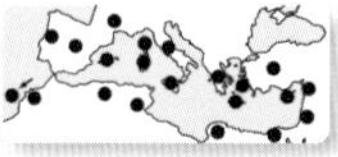

Woolly Lip Fern

Cosentinia vellea · Pteridaceae | Mar–Sep

Almost all the species occurring in the Mediterranean region are perennial, the Jersey Fern shown above being an exception. The division and nomenclature of the family has changed several times.

Identification 0.1–0.35m. Leaf stalk shiny yellowish-brown, scaly, lamina linear-lanceolate in outline, tapering to a point, bipinnate, with rounded, blunt segments, both sides covered in dense, white to yellowish, later rusty red, spreading scales under which the sori are concealed on the lower side.
Habitat Rock crevices, cracks in walls, particularly on limestone.
Other species The Cloak Fern has rough leathery leaves, their stalks shiny reddish-brown. Lamina linear-lanceolate, bipinnate, segments oblong, blunt, dark green above, with silvery white, later rusty red, spreading scales below (Mediterranean, Canaries).

Cloak Fern
Notholaena marantae

The fine, dark stems of the leaflets give the fern its name.

The dense hairs of the Woolly Lip Fern are an adaptation to dry sites.

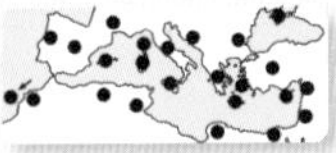

Rustyback, Scaly Spleenwort

Asplenium ceterach · Aspleniaceae | May–Aug

The Spleenworts are so named from the old belief, based on the 'doctrine of signatures', that the fern was useful for ailments of the spleen, because of the spleen-shaped sori on the backs of the fronds.

Identification Up to 0.25m. Leaves like leather, thick, with short stalk. Lamina oblong, both sides with 9–12 alternate, semicircular, simple segments, dark green above, hairless, covered beneath with dense white, later golden brown, spreading scales, standing out like cilia at the edge. Sori linear, running obliquely to the midrib, initially hidden beneath the chaffy scales.
Habitat Sunny rocks and cracks in walls on limestone.
Other species The stalk of the Irish Spleenwort is about as long as the triangular egg-shaped lamina, this 2–4 times pinnate, caudate at the end. End segments with aristate teeth. Sori close to the midrib (shady rocks, thickets, on siliceous rocks, Mediterranean, Canaries).

Irish Spleenwort
Asplenium onopteris

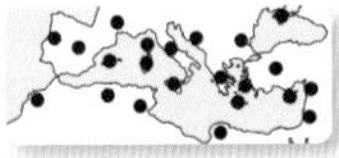

Southern Polypody

Polypodium cambricum · Polypodiaceae | Feb–Jun

The Polypodies have a creeping rhizome above ground or flat in the soil.

Identification Up to 0.5m. Rhizome with linear-lanceolate, 5–16mm long scales. Leaves produced in autumn, dying off in summer. The short-stalked, hairless, leathery lamina oblong-lanceolate, pinnately parted to almost the midrib. Leaflets narrow, serrate-crenate. Sori in 2 rows on the underside the leaflets.
Habitat Shady-damp rock crevices, tree trunks.
Other species The Mediterranean Clubmoss (Selaginellaceae) grows like a moss, pressed to the ground and freely branched. The flattened shoots with 4 rows of pairs of unequal leaflets, 2 smaller, flattened on the stem, to the side of which 1 row of larger spreading leaflets with fine (hand lens needed) saw teeth (damp shady sites, Mediterranean, Canaries).

Mediterranean Clubmoss
Selaginella denticulata

In dry summer weather the leaves roll up completely.

Polypody species have sori arranged in rows of dots on the underside of the leaf.

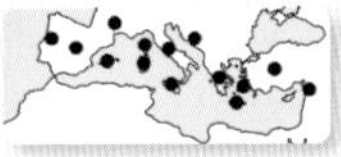

Stone Pine

Pinus pinea · Pinaceae | Apr–May

The oil-rich seeds of the Stone Pine, with their typical, nutty taste, are known as "pine nuts". Pine nuts have many uses in Mediterranean cuisine in particular, and are an essential ingredient of pesto.

Stone Pine bark

Identification Up to 30m. An easily identified pine from the umbrella-shaped crown. Trunk with greyish-brown bark, leaving reddish patches when it flakes off. Needles in pairs on short shoots, green, stiff and pointed, 1–1.5mm wide, with fine, upward pointing denticles (hand lens needed). Cones 8–15 × 7–10cm, stalkless, only ripe in 3rd year, spherical-ovate when open, shiny reddish-brown. Apophyses (outgrowths) thick, with radial ridges and large, flat, grey-white umbo (protuberance). Seeds in pairs on the cone scales, hard-coated, with early shed wing, the edible kernel 10–15mm long.
Habitat Larger, light forest stands especially in sandy areas close to the coast, timber and ornamental tree.

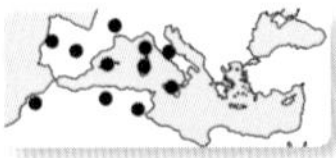

Maritime Pine

Pinus pinaster · Pinaceae | Apr–May

On the trunks of Maritime Pines you can still sometimes see a series of carved notches with a collecting container beneath. The balsam collected in this way forms the basis of the oil of turpentine used for medicines.

Male inflorescence of the Maritime Pine

Identification Up to 40m. Tree with conical crown and deeply fissured, greyish-brown bark. Needles in pairs on short shoots, shiny green, strong and sharp, 2mm wide. The shiny pale brown, later grey cones ripening in autumn of the 2nd year, stalkless, 2–8 forming a star shape, 10–22 × 5–8cm. Apophysis with pointed transverse keel and pronounced, pointed, straight or downward-pointing umbo. Seeds 7–8mm, wings up to 3cm long.
Habitat Forming forests on sandy soils and siliceous rocks, ascending up to the lower montane zone, also planted, for example for dune stabilisation.

Whether some pine stands are naturalised or not cannot be proved with certainty.

The cones of the Maritime Pine are the largest of the European pines.

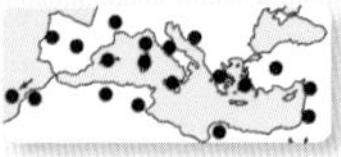

Calabrian Pine, Turkish Pine
Pinus brutia

Aleppo Pine

Pinus halepensis · Pinaceae | Mar–May

Contrary to what its name suggests, the Aleppo Pine is the commonest pine throughout the Mediterranean on limestone, forming extensive forests.

Identification 13–20m. Pine which withstands long dry periods, with light, irregular crown, branches bent or twisted. Bark silvery grey at first, smooth, later brownish. Needles in pairs on short shoots, soft and flexible, light green, only 0.7mm wide. Cones ripe in 2nd year, single or 2–3, yellow to reddish-brown, 5–12cm × 4cm, on 1–2cm long stalk. Apophysis flat, with thornless umbo. Seeds 7mm, wings 2–3cm long.
Habitat, Forming forests, often close to the coast.
Other species The Calabrian Pine has dark, stiff, 1–1.5mm wide needles. Cones 1–3, often stalkless or with 1cm long stalk, horizontal or erect (Greece and Crete to Syria, uncertain in Calabria).

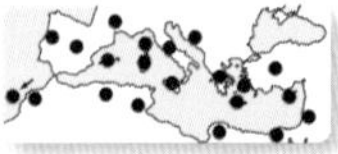

Greek Juniper
Juniperus excelsa

Phoenician Juniper

Juniperus phoenicea · Cupressaceae | Sep–Apr

In the junipers, the cone scales are fused into spherical, fleshy "berry cones" and contain 1–9 seeds. Birds eat them whole, thus dispersing the seeds.

Identification 1–2(–8)m. Monoecious shrub or small tree, on exposed coasts also prostrate. Leaves scale-like, with narrow, membranous margin, closely flattened along the branches, with a glandular furrow on the back. Leaves of young plants needle-like, spreading. Berry cones ripe in 2nd year, 8–14mm, yellowish, then brownish red, scarcely frosted, with 3–9 seeds.
Habitat Woods, garigue, mainly close to the coast.
Other species The Greek Juniper forms trees up to 20m in height. Berry cones 7–10mm, with 4–6 seeds, green at first, frosted, later dark reddish-brown or blackish (mountains, Balkan Peninsula to Syria).

The Aleppo Pine is identified by its stalked, downward-curving cones.

The inconspicuous male flowers are at the ends of the branches.

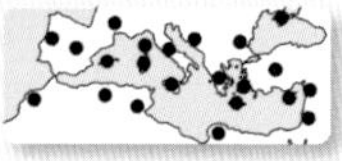

Large-fruited subspecies of the Prickly Juniper

Prickly Juniper

Juniperus oxycedrus · Cupressaceae | Oct–May

The berry cones of the Prickly Juniper are not used. It is the dark blue berry cones of the Common Juniper, *Juniperus communis*, which are used as a spice and a medicine. This montane species has a white stripe on the upper surface of the leaf.

Identification 1–8(–14)m. Dioecious shrub or small tree. Leaves in spreading whorls of 3, prickly, with 2 white stripes on the upper side. The berry cones ripen in the 2nd year, reddish-brown, with 2–3 seeds, in ssp. *oxycedrus* leaves 1–1.5mm wide, ripe berry cones 8–12mm, in the large-fruited ssp. *macrocarpa*, leaves 2–2.5mm wide, berry cones 12–15 mm, frosted.
Habitat Ssp. *oxycedrus* often in maquis and woods up into the mountains, ssp. *macrocarpa* close to the coast on sand, both throughout the Mediterranean. These are sometimes treated as separate species: *J. oxycedrus* and *J. macrocarpa*.

Funeral Cypress
Cupressus sempervirens

Funeral Cypress, Italian Cypress

Cupressus sempervirens · Cupressaceae | Jan–Apr

In the case of Funeral Cypresses, we immediately think of the tall, pillar-like trees typical of landscape paintings of Tuscany or Greece. But few know that these are cultivated forms of a spreading-branched native tree of the eastern Mediterranean.

Identification Up to 25m. Monoecious tree with ± horizontally spreading branches (f. *horizontalis*) or in the columnar form (f. *sempervirens*) with erect branches. Leaves 0.5–1.5mm, dark green, scale-like, with a gland on the back, tightly flattened on the shoots like roof tiles. Cones ripening in 2nd year, 2–3.5cm, ovate-spherical, lignified. Seeds 5–7mm, with narrow, wing-like margin. The essential oil from the leaves and young branches used primarily in cosmetics and perfumes.
Habitat Mountains, locally forming woodlands (f. *horizontalis*), also planted like the Italian Cypress.

The cones turn from green to reddish-brown as they ripen.

The cones have woody, angular, yellowish-brown to silvery-grey scales.

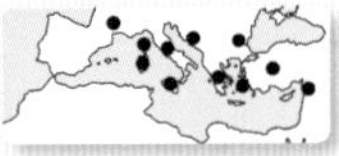

Oriental Hornbeam
Carpinus orientalis

Hop Hornbeam

Ostrya carpinifolia · Betulaceae | Apr–May

Hop Hornbeam and Oriental Hornbeam also occur together in mixed deciduous woodlands of the sub-Mediterranean Downy Oak zone.

Identification 4–10(–20)m. Deciduous shrub or small tree. Leaves egg-shaped, tapering to a point, doubly sharply serrate, becoming hairless. Leaf halves mostly unequal like the elms. Male catkins pendulous, developed in winter even before the leaves, female erect at first, forming a pendulous multiple fruit. Each nutlet completely enclosed by an inflated, egg-shaped, entire involucre.
Habitat Mixed deciduous woodland.
Other species The Oriental Hornbeam also has pendulous, cone-like multiple fruits. Nutlets with open, leaf-like, triangular-ovate, irregularly toothed involucre (Italy to Syria).

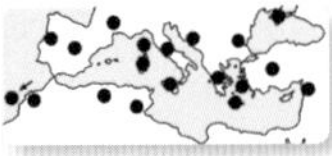

Mature fruit of the
Sweet Chestnut

Sweet Chestnut

Castanea sativa · Fagaceae | May–Aug

Eaten roasted or ground into chestnut flour, the starchy fruits once had a much greater importance as a food than they do today. A large proportion is still used as cattle feed.

Identification 10–30m. Deciduous tree, still bare in April even in S Europe. Leaves oblong-lanceolate, with thorny teeth, pale green beneath, becoming hairless, with 12–23 lateral veins. Flowers with inconspicuous perianth, male clustered in whitish, catkin-like inflorescences, with 1–3 female ones at their base with a common cupule. This densely spiny when ripe, with 1–3 dark brown fruits, dehiscing into 4 valves.
Habitat Mixed deciduous woodland in the Downy Oak zone, mostly on limestone-free soils, really only native in the eastern Mediterranean, further naturalised.

The multiple fruits are reminiscent of a hop cone.

The Sweet Chestnut is nothing like the commonly cultivated Horse Chestnut.

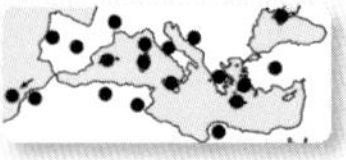

Sweet Acorn Oak
Quercus rotundifolia

Holm Oak

Quercus ilex · Fagaceae | Apr–Jun

Evergreen oaks would be the most important trees in the Mediterranean region and would cover extensive areas were it not for the human destruction of forests and soils which has been going on since ancient times.

Identification Up to 15–30m. Evergreen, monoecious tree, trunk shiny light grey, smooth, later with finely fissured bark. Leaves leathery, oblong-lanceolate, entire to toothed, dark green above, densely grey felty beneath with 7–14 pairs of lateral veins, leaf stalk up to 10mm. Male flowers in 4–7cm long catkins, female 1–2 on hairy stalks. Cupule with flattened, blunt, softly hairy scales.
Habitat Forming forests, rarer in east.
Other species The Sweet Acorn Oak has broader leaves with a blunt tip and only 5–8 pairs of lateral veins. Leaf stalk up to 6(–8)mm (western Mediterranean).

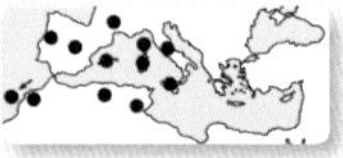

Freshly harvested cork

Cork Oak

Quercus suber · Fagaceae | Mar–Jun

The natural product cork from the bark of the Cork Oak has a wide range of uses nowadays, but as a bottle top is often replaced by plastic or metal. The first harvest after 12–15 years supplies a low-grade product, but high quality cork can then be harvested up to 10 times at intervals of 9–12 years.

Identification 10–20m. Evergreen, monoecious tree, trunk soon with exceptionally thick, corky bark. Rough, leathery, ovate-oblong leaves, shiny dark green and hairless above, weakly grey felty with raised veins beneath. Leaf margin on both sides with 4–5 sharp, short teeth or almost entire, leaf stalk 6–20mm long. Upper scales of the cupule longer and loosely incurved.
Habitat Light, evergreen woods on lime-poor soils, often planted for cork production and pig fattening.

Oaks can easily be recognised by the cupules which are typical for each species.

Freshly debarked trunks are cinnamon-brown, becoming reddish-brown to blackish later.

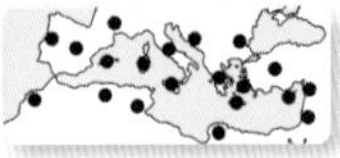

Kermes Oak

Quercus coccifera · Fagaceae | Mar–May

The Kermes Oak occurs in grazed garigue, often in the form of heavily browsed, low shrubs, in the east also as tall trees which have sometimes been distinguished as a separate species, *Quercus calliprinos*.

Identification Up to 3(–20)m. The evergreen, stiff, leathery, mature leaves hairless, dark green shiny above, paler, egg-shaped to oblong beneath. Margins sinuately-undulating, most with prickly teeth, leafstalks 2–8mm long. Cupule with short, flattened to spreading, prickly scales.
Habitat Garigue, maquis, open woods.
Other species The Valonia Oak (in this instance ssp. *macrolepis*) has larger, semi-evergreen leaves with broad teeth running into points. Upper side dull, underside felty with stellate hairs, leafstalks 5–20mm. Cupule with spreading, felty-woody scales (S Italy to Syria).

Valonia Oak
Quercus ithaburensis

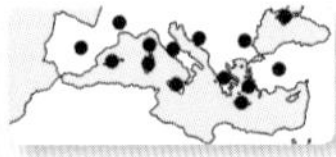

Downy Oak, White Oak

Quercus pubescens · Fagaceae | Apr–May

The sinuately lobed leaves of this characteristic tree of the evergreen, sub-Mediterranean "Downy Oak" zone are reminiscent of the oaks further north in Europe.

Identification Up to 15(–25)m. Deciduous tree with grey felty hairy buds, young twigs, leafstalks and cupules. Older leaves becoming hairless above, ± whitish beneath with felty stellate hairs, oblong egg-shaped, 4–7 times sinuately lobed on both sides and with rounded segments, leaf stalk 4–25mm.
Habitat High, light woods with abundant undergrowth.
Other species The Turkey Oak is a deciduous tree growing up to 35m in height. Leaves oblong, pinnately lobed with 4–9 pairs of irregular, blunt or pointed segments, scarcely hairy above, with ± dense stellate felty hairs beneath. Leaf stalk up to 15mm long. The awl-shaped scales of the cupule spreading (S Europe to Syria).

Turkey Oak
Quercus cerris

The hairless leaves of the Kermes Oak have spiny teeth.

The felty hairy scales lie flat against the cupule.

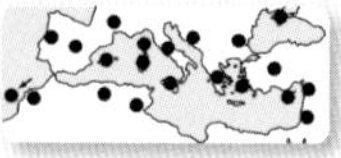

Nettle Tree, Mediterranean Hackberry

Celtis australis · Ulmaceae | Apr–May

In contrast to the winged nuts of the Elm species, the Nettle Tree has edible, fleshy, orange-yellow stone fruits which turn violet-brown.

Identification Up to 25m. Deciduous tree or shrub, leaves lopsided, narrowly egg-shaped, tapering to a long point and sharply serrate, rough above, softly-hairy beneath. Perianth reddish-brown, with 5–7 segments, falling. The sweet-tasting, fleshy, spherical, 9–12mm stone fruits on about 3cm long stalks, pendulous.
Habitat Open woods and scrub especially in the Downy Oak zone, often planted as an ornamental and street tree.
Other species The rare, deciduous Cretan Zelkova has small, dark green, egg-shaped to elliptical leaves with 6–12 coarse notched teeth. Perianth (4)5-parted. About 4mm, downy hairy, dry, lopsided stone fruits (3–5(–10)m, scrub, mountain forests, Crete).

Cretan Zelkova
Zelkova abelicea

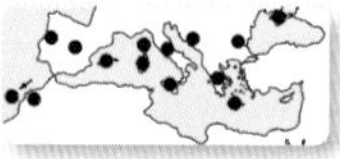

Black Mulberry

Morus nigra · Moraceae | Apr–May

The taste of the Black Mulberry is pleasantly sourish-sweet when ripe. The fruits are eaten fresh or used for making jam and syrup and also for mulberry wine.

Identification 4–10(–15)m. Deciduous cultivated tree with varied, broad ovate-heart-shaped leaves, serrate-crenate at the edge, entire or sinuately 3–5-lobed, rough above, finely hairy beneath. Flowers inconspicuous, unisexual.
Habitat Cultivated in the Mediterranean since ancient times and occasionally naturalised (native to Asia).
Other species The White Mulberry was cultivated mainly for the leaves to feed silk worms. Leaves smooth above, sparsely hairy beneath. Multiple fruit distinctly stalked, whitish, pink or purplish-violet, less flavoursome (native to China).

White Mulberry
Morus alba

Old trees also have a smooth bark like Beech.

The blackberry-like multiple fruits finally turn blackish-violet and are practically unstalked.

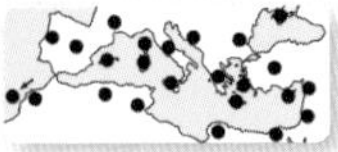

Fig

Ficus carica · Moraceae | Jun–Sep

Figs are eaten in various ways, either fresh or dried, or they are made into jam, schnapps and wine. They are also used as a gentle laxative.

Identification 2–5m. Lactiferous, deciduous tree or shrub with thick, silvery-grey branches. Leaves only produced in late spring, stalked, mostly palmately 3–5-lobed, later leathery, dark green above, rough, paler and softly hairy beneath. Flowers on the inside walls of pitcher-shaped structures with a small opening at the top. The complicated pollination in nature by gall wasps is circumvented nowadays by cultivars which produce fruits without pollination. Botanically the ripe yellow, brownish-violet or bluish-black figs are multiple fruits containing the individual fruits (nutlets) within the flesh.
Habitat Originally a plant of cliffs and rocks, cultivated tree.

Figs for sale at a southern market

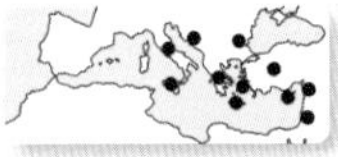

Oriental Plane

Platanus orientalis · Platanaceae | Apr–May

The Oriental Plane is a species of the riparian forest and only occurs in the eastern Mediterranean. The London Plane is more resistant to frost and is planted northwards to southern Sweden.

Identification Up to 30m. Deciduous tree with bark which flakes off in layers. Leaves long-stalked, wedge-shaped at base, 5–7 times palmately lobed to over half way, sinuately toothed. Flowers 4-parted in spherical, about 1–3cm heads, in bunches of 3–6 on a hanging peduncle. Female inflorescences with red stigmas.
Habitat Riparian forests, river banks, often along with Oleander, also planted as an ornamental or shade tree.
Other species In the London Plane the leaves are ± truncate to weakly heart-shaped at base, 3–5-lobed to the middle at most, sharply toothed. Flower heads mostly 2 (cultivated plant).

London Plane
Platanus hispanica

The flowers, hidden in pitcher-shaped structures, appear before the leaves.

Female and male flowers grow on separate heads.

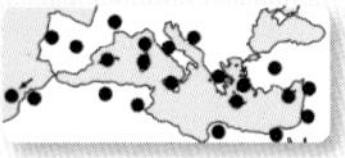

Carob, St. John's-bread, Locust Bean

Ceratonia siliqua · Caesalpiniaceae | Aug–Jan

The dried seeds of the Carob were formerly used as troy weight (carat) due to their constant weight (0.18g). Locust Bean gum has a high swelling capacity and is used as a thickener and stabiliser in food.

Fruits and female flowers

Identification 2–15m. Evergreen shrub or tree with spreading branches. Leaves even-pinnate with 4–10 short-stalked, reverse egg-shaped, leathery leaflets, dark green above. Flowers inconspicuous, in spreading racemes. Pods up to 25cm long, flat, dark reddish-brown, leathery, with 10–17 seeds. Nowadays the fruits (carob bean pods) containing up to 40% sugar are predominantly used as cattle feed and for the extraction of alcohol. They are also edible for humans.
Habitat Maquis close to the coast, especially in the southern Mediterranean. Cultivated tree in many varieties.

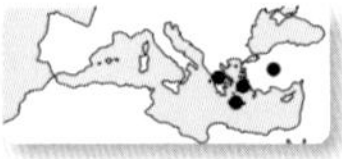

Cretan Maple

Acer sempervirens · Aceraceae | Apr–May

Mediterranean maple species, like those from further north in Europe, can be recognised by their fruits composed of two fruitlets, each with a one-sided, membranous, prominently net-veined wing.

Montpellier Maple
Acer monspessulanum

Identification 1–5(–12)m. Evergreen, semi-evergreen or deciduous shrub or tree. Leaf stalk 0.5–1.5cm long, lamina leathery, mostly 3-lobed, with tough edge, undulating crenate-toothed, hairless on both sides. Few greenish-yellow, inconspicuous, 5-parted flowers on short stalks in small corymbs.
Habitat Rocky limestone slopes, montane forests.
Other species The Montpellier Maple is deciduous. Leaf stalk 2–6cm long, lamina entire, 3-lobed, initially softly hairy beneath. Flowers long-stalked, fruits with almost parallel wings (deciduous woods, thickets, Mediterranean, rarely as far as the warmest parts of Central Europe).

The inflorescences grow directly on the trunk and branches (cauliflory).

The expanding fruit wings are often bright red.

Cretan Date Palm

Phoenix theophrasti · Palmae | Feb–Jun

The Cretan Date Palm was first described in 1967 and separated from the Date Palm. Some stands are now protected.

Identification 1.5–10m. Dioecious trees, multiple stemmed, trunk up to 35cm across, not over 10m high. Leaves grey-green, curved erect, later pendulous, pinnatisect, the central segments 30–50cm long, lower shorter and thorny, yellowish. Inflorescences initially sheathed by bracts, much branched. Fruits approx. 1.5cm, yellow-brown at first, blackish when ripe, fibrous, non-edible.
Habitat Sandy river valleys and river mouths, rocky limestone areas with groundwater.
Other species The cultivated Date Palm has only 1 stem and is up to 30m high. Fruits 2.5–7.5cm, oblong, yellowish-orange to brownish-red, fleshy, palatable, often protected with film before the harvest (irrigated crop in N Africa, SW Asia, S Spain (Elche), ornamental plant).

Date Palm
Phoenix dactylifera

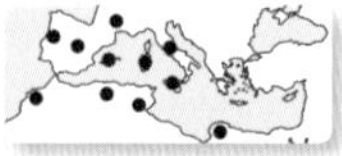

European, Mediterranean or Dwarf Fan Palm

Chamaerops humilis · Palmae | Mar–Jun

Besides the Cretan Date Palm *Phoenix theophrasti*, the European Fan Palm is the second palm native to Europe. Often bushy and ± stemless due to grazing, the species only produces a trunk in protected sites or cultivation.

Identification 0.5–4m. Leaves in terminal tuft, green to grey-green, lamina 70–80cm across, rounded, up to two-thirds divided into 10–20 lance-shaped pointed segments forming a fan, the leaf stalk with thorny teeth at the edge. Numerous flowers in sheathed, dense, yellow, paniculate inflorescences. Fruits 1–3cm, spherical or oblong, yellow, later reddish brown, non-edible. The young palm shoots are eaten as palmitos.
Habitat Rocky ground and on sand, often forming stands; only in local, small stands at the limit of distribution in Italy, common as an ornamental plant.

Inflorescences of the
European Fan Palm

The Cretan Date Palm grows with multiple stems from shoots.

The European Fan Palm often forms a distinct garigue type over large areas.

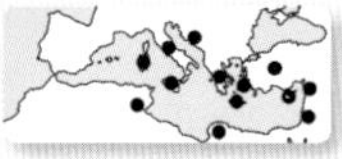

Thorny Burnet

Sarcopoterium spinosum · Rosaceae | Feb–May

The plant has been used since ancient times to fence off gardens and farms to keep grazing cattle out. As it is abundant around Jerusalem, it is one of the thorny shrubs said to have been used for Christ's crown of thorns.

Identification 0.3–0.6m. Dense, squarrosely branched dwarf shrub with leafless, angularly branched, thorny shoots. Leaves with 9–17 oblong-ovate, incised serrate, hairy leaflets in the top part. Flowers without petals with separate sexes in terminal, oblong heads ± stalkless, the upper female with red, feathery stigmas, the lower male with 10–30 long, yellow stamens. Calyx 4-fold, expanding after the flower and attached to the reddish, spherical, 3–5mm fruits.
Habitat Garigue, often in large stands.

Male flowers of the Thorny Burnet

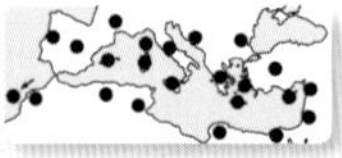

Castor Oil Plant

Ricinus communis · Euphorbiaceae | Feb–Oct

Harvested from the seeds and freed from the highly toxic protein ricin, castor oil is used as a laxative and as an ingredient in cosmetics. Larger quantities are now used technically as lubricants and in the manufacture of plastics.

Identification 0.5–4m. Hairless, fast-growing plant up to tree height, with large, long-stalked leaves. Flowers dioecious in erect panicles, the male with branched, yellow stamens, the female above with conspicuous, red stigmas. The capsule mostly softly spiny and with 3 bean-like, shiny, reddish-brown and greyish-white marbled, highly poisonous seeds, eating only a few of which can be fatal!
Habitat Roadsides, ruderal areas, also ornamental plant. Native of tropical Africa, where also a crop plant.

Castor Oil Plant seeds

The female flowers stand out because of their feathery red stigmas.

Castor Oil Plant leaves are 5–9 times palmately lobed and toothed.

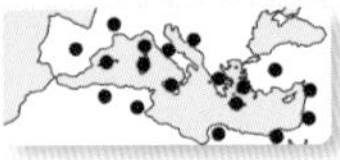

Tree Spurge

Euphorbia dendroides · Euphorbiaceae | Nov–Apr

The distinguishing feature of spurges is the cyathia with a cup-shaped involucre with 4 or 5 nectaries on the margin. They surround 5 groups of male flowers, each containing one stamen and a central, stalked, female flower. Almost all spurges have poisonous milky sap.

Cyanthia of the Tree Spurge

Identification 0.5–2(–3)m. Tall, summer-deciduous, hairless spherical bush, with regular forked branching and stem often as thick as an arm. Leaves concentrated at the ends of the stems, oblong-lance-shaped, often with distinct point. The umbel-like inflorescence with 4–10 rays, the rays also with forked branches, each with 2 free, yellowish bracts below the cyanthia. Nectaries transversely elliptical, ± lobed. Capsule with 3 furrows, smooth like the grey seeds.
Habitat Mostly on limestone rocks close to the coast, locally often in larger stands.

Spiny Spurge

Euphorbia spinosa · Euphorbiaceae | Apr–Jun

The two following spurge species become thorny shrubs because the dead twigs and umbel rays persist at first.

Greek Spiny Spurge
E. acanthothamnos

Identification 0.1–0.5m. Cushion-forming, much branched, hairless shrub with rigid branches and umbel rays from the previous year. Leaves blue-green, lance-shaped, much wider at the base of the cymose umbel. Umbel rays 1–5, short, each with 1 cyanthium, nectaries oval. Capsule often with long, conical warts. Seeds smooth, brown.
Habitat Garigue, rocks, ascending to the montane zone, calcicolous.
Other species The distribution of the similar Greek Spiny Spurge continues eastwards. Umbel rays mostly 3, often forked 2–3 times, prickly. Stem and involucral bracts elliptical, similar, bright green. Capsule with short, cylindrical warts (Balkan Peninsula, Crete, W Anatolia).

Leaves and fruits turn orange to red in summer and are shed.

The twigs and umbel rays of the previous year are rigid but not spiny.

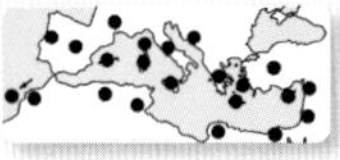

Mastic Tree, Lentisc

Pistacia lentiscus · Anacardiaceae | Mar–Jun

Mastic is the name given to the glass-like resin obtained from incisions in the bark, with the best quality coming from cultivated trees especially on the Island of Chios. Its uses include chewing-gum, securing wound dressings and in scented incense blends.

Identification 1–3(–8)m. Shrub, more rarely a small tree, dark green even in the dry season. Leaves even-pinnate, the 8–12 elliptical-lanceolate, blunt leaflets with small, distinct tip and winged rachis. Wind-pollinated dioecious flowers without petals in short, dense inflorescences in the leaf axils, male with red anthers, female greenish. Stone fruits about 4mm, red, later black.
Habitat Abundant species of the maquis and open woods throughout the Mediterranean region.

Mastic, resin from the Mastic Tree

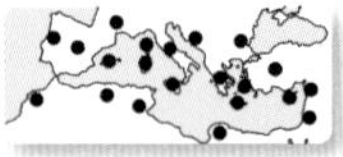

Turpentine Tree, Terebinth

Pistacia terebinthus · Anacardiaceae | Apr–Jul

Pistachios are familiar to everyone as edible "seeds". However, these do not come from the Mediterranean species, but from the Common Pistachio which is also cultivated in the eastern Mediterranean.

Identification 2–5m. Deciduous, aromatic shrub or small tree. Leaves odd-pinnate, with 3–9 leaflets. Brownish, dioecious flowers without petals, in long panicles. Stone fruits 5–7mm, spherical-ovate, pink, bluish, later brownish. The terminal leaflet in the eastern Mediterranean subspecies is lacking or very small.
Habitat Maquis, woods, up to the montane zone.
Other species The Common Pistachio has rounded, thin, (1–)3–5-parted leaves. Pink, 2–3cm stone fruits with pale, hard, opening stone which contains the edible, greening seed leaves in a reddish-brown seed coat, (native of Central Asia).

Common Pistachio
Pistacia vera

The anthers of the male flowers are bright red when closed.

The pointed, oval leaflets are attached to a wingless rachis.

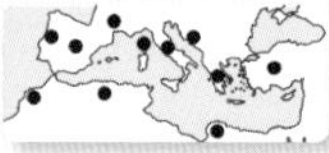

Box

Buxus sempervirens · Buxaceae | Feb–May

The Box has been a favourite garden plant in Europe in many varieties since ancient times. Its wild form can often be found in the Mediterranean region and sometimes further north as a tall shrub.

Identification 0.5–3(–7)m. Poisonous, evergreen shrub. The opposite, short-stalked, leathery leaves entire, ovate-elliptical, 1.5–3cm long. Flowers in a 5–7mm wide cyme with stalkless flowers formed from a terminal, 4–6-parted, whitish female flower and several 4-parted, greenish-yellow male ones. The perennial style less than a quarter as long as the leathery capsule which becomes blackish-brown.
Habitat Evergreen and deciduous broad-leaved forests.
Other species The Balearic Box has larger, 2.5–5cm long leaves. Cyme with sessile flowers 7–10mm across. Style over a quarter as long as the capsule (Sardinia, Balearics, occasional in Spain, NW Africa, S Anatolia).

Balearic Box
Buxus balearica

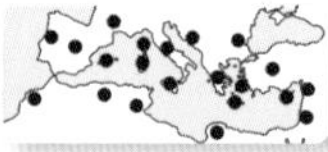

Mock Privet, Green Olive Tree

Phillyrea latifolia · Oleaceae | Mar–Jun

With their tough, entire leaves, the *Phillyrea* belong to the typical species of the evergreen woods and maquis of the Mediterranean.

Identification 2–8m. Evergreen shrub or small tree. Leaves opposite and hairless, leathery, egg-shaped and mostly distinctly toothed, 8–40mm wide, with conspicuous lateral veins. Adult leaves narrower, less toothed. Flowers in small racemes, greenish white, with 4-lobed corolla. Calyx incised up to three quarters of the length. 4–7mm, spherical, later bluish-black stone fruits.
Habitat Maquis, open woods.
Other species The Narrow-leaved Mock Privet is similar. Leaves all similar, linear-lanceolate, 3–12mm wide, mostly entire. Calyx incised to about a quarter of the length. Intermediate forms of both species have been described (central and western Mediterranean, Canaries).

Narrow-leaved Mock Privet
P. angustifolia

The cyme of sessile flowers consists of one female and several male flowers.

The leaves are variable, from thorny toothed to almost entire.

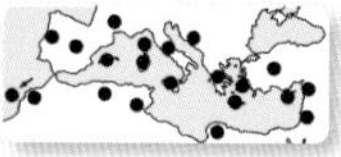

Large-leaved Nettle

Urtica membranacea · Urticaceae | Feb–Sep

Besides the Common or Stinging Nettle and the Small Nettle, which are also widespread further north in Europe, some interesting species occur in the south, the two most abundant of which are described here.

Identification 0.2–1.2m. Annual, mostly dioecious plant with stinging hairs. Leaves opposite, egg-shaped acute, coarsely serrate, with 4 stipules joined in pairs. Single sex inflorescences in the axils, long male ones with an expanded axis, with flowers attached on one side, female ones much shorter.
Habitat Shady, nitrogen-rich ruderal sites.
Other species The Roman Nettle has 4 stipules at each node. Male and female inflorescences in the same leaf axils, the female in pill-shaped, spherical heads. Formerly cultivated for the viscous oily seeds (Mediterranean).

Roman Nettle
Urtica pilulifera

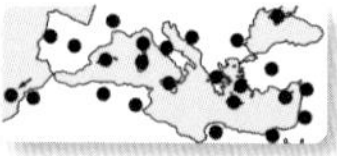

Pellitory-of-the-Wall

Parietaria judaica · Urticaceae | Apr–Sep

Parietaria **species prefer shady sites on walls or in wall crevices. The name comes from Latin** ***paries*****: wall.**

Identification 0.1–0.6m. Perennial, branched at base, ± erect, softly hairy plant. Leaves 2–7cm long, alternate, pointed egg-shaped, narrowed at base, without stipules. Several inconspicuous, 4-parted flowers in the axils, their bracts joined at the base, shorter than the perianth at fruiting time. Small, black, shiny nuts.
Habitat Mostly damp, shaded walls.
Other species The small Cretan Pellitory-of-the-Wall has a prostrate to ascending growth form. Leaves broadly egg-shaped, usually only up to 1.5cm. Bracts at fruiting time expanded, brown, forming an unequal 5-lobed involucre around the multiple fruit, nuts brown (Sardinia, eastwards to Asia Minor).

Cretan Pellitory-of-the-Wall
Parietaria cretica

The female inflorescences are much shorter and not expanded.

Parietaria species belong to the Urticaceae, but do not have stinging hairs.

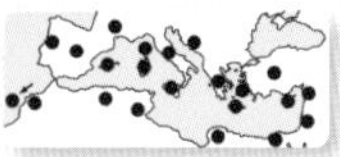

Horned Dock, Red Dock

Rumex bucephalophorus · Polygonaceae | Mar–Sep

The annual Horned Dock with the unwieldy scientific name (meaning "ox-head carrying") is widespread in open sites and very variable.

Identification 0.1–0.4m. Stems single, erect and sturdy, or several, branched at the base and ascending. Leaves narrow spatulate or ovate-lanceolate. Flowers usually 2–3, their stalks bent downwards, sometimes widened into a club shape. At fruiting time the inner of the 6 perianth segments very enlarged, with distinct teeth on both sides and a small callus ridge.
Habitat Cultivated land, waste ground, on sandy soils.
Other species The Spiny Dock is considerably sturdier. Female flowers with 6 joined tepals, of which the 3 outer ones expanded with spreading thorns at fruiting time (up to 0.6m, Dec–Jun, Mediterranean).

Spiny Dock
Emex spinosa

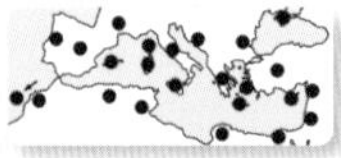

Prickly Saltwort

Salsola kali · Amaranthaceae | Jul–Oct

***Salsola* species contain a lot of alkali salts, so that their ash was formerly used as a washing agent. The young leaves are used in salads or as a vegetable but their poisonous effects are the subject of debate.**

Identification 0.1–1m. Annual, variable, spreading branched, fleshy, hairless or bristly hairy plant. Only the lower leaves opposite, linear-awl-shaped with spiny tip, widened and with a membranous margin at the base. Flowers 1–3 in the leaf axils, their 5 sepals unequally wide, with a tubercle or membranous wing at the back, exceeded by a leaf and 2 stiff, thorny secondary bracts.
Habitat Sandy coasts, salty sites inland.
Other species The Opposite-leaved Saltwort is hairless. The fleshy leaves with short, soft point, opposite to far along the stem, hemicylindrical, almost clasping the stem. Secondary bracts pointed, without thorns (Mediterranean).

Opposite-leaved
Saltwort
Salsola soda

The whole plant is often tinged red.

Prickly Saltwort is typical of coastal drift lines.

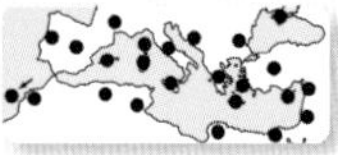

Sea Spurge

Euphorbia paralias · Euphorbiaceae | May–Nov

Two very different Spurge species occur on unlevelled sandy beaches in the Mediterranean. Both are typical of this site.

Identification 0.2–0.7m. Perennial, hairless, stiffly erect, grey-green and somewhat fleshy spurge. The numerous leaves arranged like roof tiles, pointing upwards in the upper section, oblong-elliptical, the topmost egg-shaped. Cyanthium with 3–6 rays, forked up to 3 times. Floral bracts free, kidney-shaped. Nectaries notched at tip, with short horns. Capsule hairless, with smooth seeds.
Habitat Sand and shingle beaches.
Other species The annual Purple Spurge has prostrate branches with characteristic asymmetrical egg-shaped to oblong-sickle-shaped leaves. Cyanthia single in the leaf axils (Mediterranean, Canaries).

Purple Spurge
E. peplis

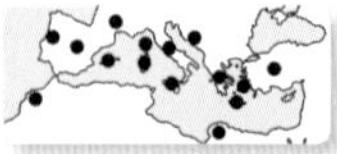

Mediterranean Spurge

Euphorbia characias · Euphorbiaceae | Feb–Jul

The stately Mediterranean Spurge can be divided into two subspecies which may also be viewed as distinct species. The centre of distribution of ssp. *characias* is in the west, that of ssp. *wulfenii* more towards the east.

Identification 0.3–0.8(–1.8)m. Perennial, hairy spurge, woody at base. Leaves in upper part of the robust stem crowded, linear to lance-shaped. The terminal cymose umbel with 10–20(–40) rays, with up to 40 rays in the leaf axils beneath. Floral bracts egg-shaped, up to a third to two-thirds joined. Capsule hairy, seeds smooth, grey.
Habitat Maquis, open woods, grazed land.
Other species The Serrated Spurge is unmistakable: upper leaves light green, with wide egg-shaped base and distinctly irregular teeth at the edge. Cymose umbel with 3–5 rays (western Mediterranean, Canaries).

Serrated Spurge,
Sawtooth Spurge
Euphorbia serrata

Below the terminal cyme are up to nine side branches.

Ssp. *wulfenii* has yellowish nectaries, ssp. *characias* reddish-brown ones (top right).

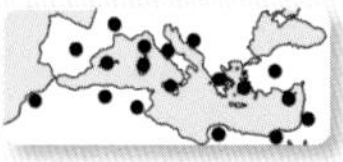

Plantago crassifolia

Plantago crassifolia · Plantaginaceae | May–Jun

The majority of the roughly 60 species of plantain which grow in the Mediterranean region have their basal leaves arranged in a rosette. The inconspicuous, 4-parted flowers form spikes in the axils of small bracts on a leafless stalk.

Identification 0.1–0.4m. Perennial plantain with several rosettes of fleshy, mostly hairless, channelled linear, sometimes distally minutely toothed, non-rigid leaves. The cylindrical, dense flower spikes 4–6cm long. Bracts acute egg-shaped, about as long as the sepals, exceeded by the finely hairy flower tube.
Habitat Coasts and other saline sites.
Other species The annual Hare's-foot Plantain is similar to the Ribwort Plantain found further north. A notable feature of the inflorescence is the long silky hair on the bracts and sepals (Mediterranean, Canaries).

Hare's-foot Plantain
Plantago lagopus

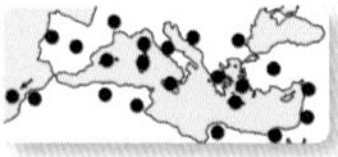

Glandular plantain

Plantago afra · Plantaginaceae | Mar–Jul

Some plantain species do not produce a leaf rosette but have opposite leaves on the stem and little branches in the axils. They were previously also separated off in their own genus *Psyllium*.

Identification 0.1–0.4m. Annual, mostly erect plant, glandular-downy above. Leaves hairy, stalkless, linear-lanceolate, rarely distally toothed. Inconspicuous membranous flowers of 4-parted corolla and calyx. Bracts all equal, without lateral veins. The pretty, dark red-brown, boat-shaped seeds are valued as a laxative due to their high mucous content and therefore their capacity to swell.
Habitat Fields, waste ground, roadsides, garigue.
Other species The much branched Shrubby Plantain is woody at the base, the awl-shaped points of the lower bracts elongated and spreading (SW Europe).

Shrubby Plantain
Plantago sempervirens

The sturdy, flattened hairy stalks extend well beyond the leaves.

The head-like flower spikes are borne on stalks from the upper leaf axils.

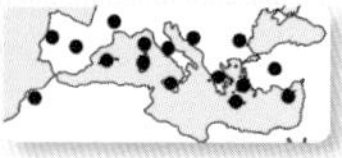

Wall Pennywort, Navelwort

Umbilicus rupestris · Crassulaceae | Apr–Jun

Umbilicus **species have rounded shield-shaped basal leaves. Their stalk is attached in the centre of the leaf on the underside and sunken like a navel on the upper side, giving the plant one of its names.**

Horizontal Navelwort
Umbilicus horizontalis

Identification 0.2–0.5m. Perennial plant with 1.5–7cm wide basal leaves, crenate at the edge, stem leaves kidney-shaped to linear. Stem to at least two-thirds of its length with dense 2–5mm long stalked, hanging flowers. Corolla whitish to greenish, often tinged red, rarely completely red, tubular, 6–9mm long, with 5 short lobes.
Habitat Shady damp rocks, cliffs and walls.
Other species The Horizontal Navelwort has ± unstalked, horizontally spreading flowers. Corolla pitcher-shaped, 4–7mm long (central and eastern Mediterranean). Replaced on the Balearics, in Spain and Morocco and in the Canaries by the similar *Umbilicus gaditanus*.

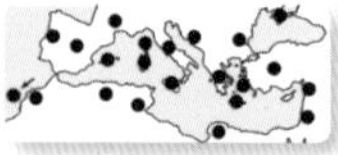

Rough Cocklebur

Xanthium orientale · Compositae | Jul–Sep

Xanthium **species are initially hard to recognise as belonging to the Compositae. The inconspicuous flowers are borne in unisexual heads, the female are 2-flowered and sunk into the egg-shaped receptacle.**

Spiny Cocklebur
Xanthium spinosum

Identification 0.2–1m. Annual, aromatic plant, branched from the ground upwards, stem often with violet mottling. Leaves long-stalked, broad ovate triangular, heart-shaped or wedge-shaped at base, entire or 3–5-lobed, toothed. In the subspecies ssp. *italicum* common in the area, fruiting heads 1.5–3.5cm long, covered with dense thorns.
Habitat Ruderal areas, especially on sand, river banks, roadsides. Origin uncertain.
Other species The Spiny Cocklebur has unstalked to short-stalked leaves, dark green with grey veining above, grey felty beneath; at the base of the leaf stalk are vigorous, 3-fold, straw-coloured thorns (native to South America).

Navelworts are hairless, fleshy plants of rocks and cliffs.

The fruiting heads are covered in hooked thorns and end in two beaks.

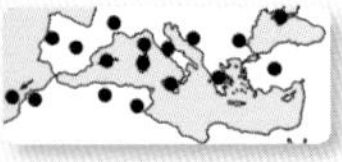

Italian Lords-and-Ladies

Arum italicum · Araceae | Apr–Jun

The Araceae are easy to recognise from the spathe and spadix. Only the Italian Lords-and-Ladies is widespread in the Mediterranean region, but there are also several local *Arum* species in the east.

Identification 0.2–0.7m. Perennial plant with horizontal tuber. Leaves appearing in autumn, long-stalked, with arrow-shaped or spear-shaped, usually light-veined lamina. Floral bract pale greenish white or yellowish. Topmost part of spadix yellow and naked, with the male flowers, framed by infertile ones and the female flowers below. Berries red. Poisonous.
Habitat Thickets, hedges, plantations.
Other species In the Cretan Arum the light yellow spathe is recurved and often twisted at flowering time, the deeper yellow spadix projecting (Crete to SW Turkey).

Cretan Arum
Arum creticum

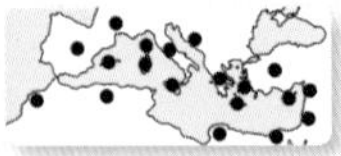

Friar's Cowl

Arisarum vulgare · Araceae | Oct–Apr

The Friar's Cowl gets its English name from the forward bending spathe tube, the spadix is curved downwards at the end.

Identification 0.1–0.4m. Perennial plant with fleshy rhizome. Leaves basal, ovate-arrow-shaped. Spathe 3–5cm long, with the upper free part hood-shaped and covering the projecting naked part of the spadix. The latter with 4–6 female flowers at the base and about 20 male flowers above these, without infertile flowers. Green Berries.
Habitat Cultivated land, waste ground, garigue.
Other species The Dragon Arum has foot-shaped leaves divided into 9–15 segments with a conspicuous spotted sheath. The inside of the up to 50cm long spathe and the spadix dark purple, with a smell of carrion (up to 1.2m, nutrient-rich sites, France to Turkey).

Dragon Arum
Dracunculus vulgaris

The erect spathe is more than twice as long as the spadix.

The spathe is fused into a green or reddish-brown striped tube.

Neptune Grass

Posidonia oceanica · Posidoniaceae | Oct–May

On beaches you sometimes find an abundance of washed-up brown "sea balls": leaves ripped off by the waves, ground and rolled into balls up to fist-sized, plus the rhizomes with fibrous leaf remains.

"Sea balls" of Neptune Grass

Identification Up to 0.5m. Perennial, submerged water plant with sturdy rhizome with the brown, fibrous remains of dead leaves. At the end 5–10 up to 55cm long, dark green, ribbon-like, 6–10mm wide, 13–17-veined leaves, rounded at the top. Flowers rarely forming, without perianth, in long-stalked inflorescence composed of spikes and surrounded by 2 leaf-like floral bracts. Stone fruit dispersed by floating in the water.

Habitat On fine sandy substrate with water depth of 0.5–45 m. Locally also on the Atlantic coast of SW Europe.

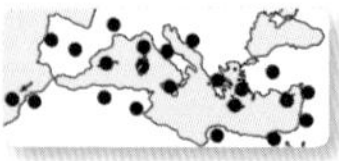

Sharp Rush

Juncus acutus · Juncaceae | Apr–Jul

The Sharp Rush lives up to its name as it does not just have an impressive form which makes it easy to identify, but can also give painful stabs.

Sea Rush
Juncus maritimus

Identification 0.3–1.5m. Perennial, large rush forming dense tussocks with stiff, prickly, cylindrical leaves. Flowering shoots 2–5mm wide, often exceeded by the lowest, stem-like subtending bract. Inflorescence many-flowered, usually dense, involucral bracts 6, about equally long, with membranous margin, the 3 inner broader, with membranous auricles at the top.

Habitat Sandy coasts, saline and freshwater marshes.

Other species With its creeping rhizome the Sea Rush forms rows or swards rather than tussocks. Leaves less sharply pointed, inflorescence loose, involucral bracts straw yellow, of unequal length. Capsule as long as the perianth or somewhat longer (Mediterranean, Canaries).

The remains of Neptune Grass can be uncomfortable for beach users.

The shiny red-brown capsules are up to twice as long as the perianth.

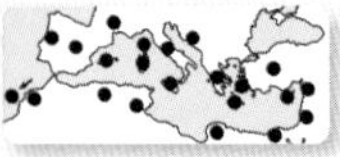

Golden Dog's-tail

Lamarckia aurea · Gramineae | Mar–Jul

The delicate Golden Dog's-tail got its scientific name from the famous French biologist Jean-Baptiste de Lamarck. The genus only contains this one species.

Identification 0.05–0.25m. Annual grass with pale green, soft, flat leaves, ligule 5–10mm long, the topmost sheath somewhat inflated. The oblong-oval panicle with secund spreading, fertile and non-fertile spikelets on hairy petiolules, the fertile ones with awned lemmas.
Habitat Roadsides, waste ground, walls.
Other species The Rough Dog's-tail has dense, egg-shaped and secund, compressed spike-like panicles. Fertile and non-fertile spikelets with long-awned glumes (0.1–1m, grassy areas, open woods, also an ornamental grass, Mediterranean, Canaries).

Rough Dog's-tail
Cynosurus echinatus

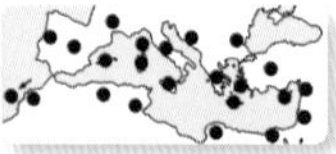

Greater Quaking Grass

Briza maxima · Gramineae | Apr–Jun

The spikelets hanging on 6–20mm long stalks as fine as hairs move in the slightest breath of wind.

Identification 0.1–0.6m. Annual grass forming loose tussocks or growing singly erect. Leaves 3–8mm wide, flat and thin, slightly rough at the edges. Ligule rounded, 2–5mm. Inflorescence in a nodding panicle of 1–12 ovate-heart-shaped, laterally compressed, 14–25mm long, 7–20-flowered spikelets. Husks almost horizontally spreading from the spikelet axis, the glumes rounded at the end.
Habitat Garigue, pastures, cultivated land, roadsides, also cultivated as an ornamental plant, in dried flower arrangements.
Other species The Lesser Quaking Grass much smaller 3–5mm, 4–8-flowered spikelets, ligule acute, 3–8mm long (damp grassy areas, ruderal sites, Mediterranean, Canaries).

Lesser Quaking Grass
Briza minor

Golden Dog's-tail stands out due to its golden yellow shimmering inflorescences.

The glumes of the often reddish-brown or violet-tinged spikelets are without awns.

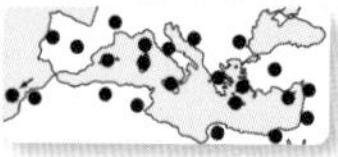

Marram Grass

Ammophila arenaria · Gramineae | May–Aug

Two typical grasses occur on the dunes of all the sandy beaches of the Mediterranean (at least if they have not become victims of tourism): Marram Grass on the main dunes, Sand Couch also on the foredunes.

Identification 0.5–1.2m. Perennial, tussock-forming grass. Leaves grey-green, 2–5mm wide, stiff, inrolled upwards, finely hairy on the ribs of the upper side. Ligules 2-pointed, 1–3cm long. Lemma with 4–6mm long, white, fine hairs at base.
Habitat Ssp. *arundinacea* along the coasts of the Mediterranean. Also planted for dune stabilisation.
Other species The Sand Couch forms loose swards. Inflorescence a spike with 8–12 distal, compressed, awnless, 5–9-flowered spikelets arranged in 2 rows (Mediterranean).

Sand Couch
Elytrigia juncea

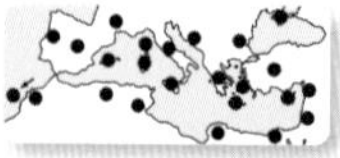

Hare's-tail

Lagurus ovatus · Gramineae | Apr–Jul

The softly hairy Hare's-tail is familiar as a component of dried flower arrangements.

Identification 0.1–0.5m. Annual, ± erect grass with flat leaves, silkily hairy on both sides, upper leaf sheath inflated, ligule 3mm, blunt, notched. The softly hairy, egg-shaped, 1–6cm long inflorescences of 1-flowered, very short-stalked spikelets are characteristic.
Habitat Sandy soils close to the coast, waste ground, garigue, pinewoods.
Other species The Ovate Goatgrass has an egg-shaped, 1–2 cm long flowering spike with 2–3 fertile spikelets on a flat, broad rachis, with 1 infertile spikelet above and 1–2 stunted ones below. Glumes of the lower fertile flower with 4–6 bristly, 2–3cm long awns (Mediterranean, Canaries).

Ovate Goatgrass
Aegilops geniculata

The long, dense panicles contain single-flowered, almost awnless spikelets.

The long awns of the lemmas project beyond the egg-shaped inflorescence.

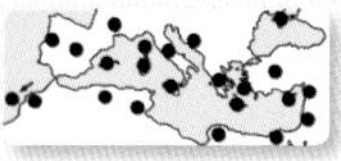

Giant Reed, Spanish Cane

Arundo donax · Gramineae | Aug–Dec

The largest European grass is reminiscent of bamboo with its thick, woody stems. It has many uses, for example for windbreaks or supports in horticulture, roofing material to create shade, and for mouth pieces of reed instruments such as the clarinet.

Identification 2–4(–8)m. Perennial grass with thick rhizome. Stalks 1–3(–5)cm wide. Leaves grey-green, flat, up to 6cm wide, with rough edges and upper sides. Flowers in 30–70cm long, dense, contracted panicles. Spikelets 3–6-flowered, often violet tinged.
Habitat Ditches, river banks, ponds, long in cultivation and naturalised, probably native to Central Asia.
Other species Stramma or Dis Grass grows in large tussocks. Leaves tough, very rough and strongly ribbed, later inrolled. Flowering panicles loose, ± on one side (1–3m, garigue, maquis, S Europe to the Aegean, NW Africa).

Stramma, Dis Grass
Ampelodesmos mauritanicus

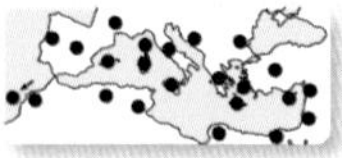

Capitate Galingale

Cyperus capitatus · Cyperaceae | Apr–Sep

The Capitate Galingale forms large, open swards by means of its creeping rhizomes, thus also helping dune stabilisation.

Identification 0.1–0.5m. Perennial beach plant. Stems single, rounded and grooved, with 1–6 mm wide, channelled, grey-green leaves at the base. Inflorescence terminal, 15–30mm across, composed of contracted head-like, 4–12-flowered, compressed spikelets. Glumes reddish-brown, with spiny tip.
Habitat Sandy coasts.
Other species The imposing, 2–5m tall Papyrus has leafless, triangular stems. Up to 100 flowering spikes on long, thin rays (2–5m, slow-flowing water bodies, probably naturalised on Sicily and Malta, and in Israel, Egypt, additionally as an ornamental plant, formerly for manufacturing "papyrus" paper, native of eastern tropical Africa).

Papyrus
Cyperus papyrus

The long, silkily hairy lemmas give the inflorescences a silvery sheen.

The inflorescence is surrounded by three bracts that are bent downwards.

Further Reading

ACEBES GINOVÉS, J. R. et al. Pteridophyta, Spermatophyta in: Lista de especies silvestres de Canarias. Gobierno de Canarias 2009

ALI, S. I., S. M. H. Jafri & A. El-Gadi (Hrsg.): Flora of Libya.

BAUMANN, H.: Griechische Pflanzenwelt in der Antike, 2. Kilchberg 2011

BAUMANN, H., S. Künkele & R. Lorenz: Orchideen Europas. Stuttgart 2006

BLANCA, G. & al. (Hrsg.): Flora Vascular de Andalucía Oriental, 4 Vols. Sevilla 2009

BOLÓS, O. de & J. Vigo: Flora dels Països Catalans, 4 Vols. Barcelona 1984–2001

BOULÓS, L.: Flora of Egypt, 4 Vols. Cairo 1995–2005

CASTROVIEJO, S. & al. (Hrsg.): Flora Iberica, Vols 1–8, 10–15, 17, 18, 21. Madrid 1986–2012

CONTI, F. et al.: An Annotated Checklist of the Italian Vascular Flora. Rome 2005

CUÉNOD, A., G. Pottier-Alapetite & A. Labbe: Flore analytique et synoptique de la Tunisie. Cryptogames vasculaires, Gymnospermes et Monocotyledones. Tunis 1954

DOMAC, R.: Flora Hrvatske. Zagreb 2002

DAVIS, P. H.: Flora of Turkey and the East Aegean Islands, 11 Vols. Edinburgh 1965–2002

DELFORGE, P.: Orchids of Europe, North Africa and the Middle East. London 2006

DÜLL, R. & I.: Taschenlexikon der Mittelmeerflora. Wiebelsheim 2007

GUINOCHET, M. & R. de Vilmorin: Flore de France, 5 Vols. Paris 1973–1984

HEGI, G.: Illustrierte Flora von Mitteleuropa, Vols 1–7, 1–3. München, Berlin, Jena 1906–2007

JAHN, R. & P. Schönfelder: Exkursionsflora für Kreta. Stuttgart 1994

JEANMONOD, D. & J. Gamisans: Flora Corsica. Aix-en-Provence 2007

KRETZSCHMAR, H., W. Eccarius & H. Dietrich: Die Orchideengattungen Anacamptis, Orchis, Neotinea. Bürgel 2007

LAFRANCHIS, T. & G. Sfikas: Flowers of Greece. 2 Vols. Paris 2009

LÓPEZ GONZÁLEZ, G. Los árboles y arbustos de la Península Ibérica e Islas Baleares, 2 Vols. Madrid, Barcelona, Mexico 2001

MAIRE, R.: Flore de l'Afrique du Nord, 16 Vols. Paris 1952–1987

MEIKLE, R. D.: Flora of Cyprus, 2 Vols. Kew 1985

MOUTERDE, R: Nouvelle flore du Liban et de la Syrie, 3 Text- and 3 Atlas vols. Beirut 1966–1986

PIGNATTI, S.: Flora d'Italia, 3 Vols. Bologna 1982

PILS, G.: Flowers of Turkey. A photo guide. Linz 2006

POTTIER-ALAPETITE, G.: Flore de la Tunisie, Angiospermes, Dicotyledones, 2 Vols. Tunis 1979, 1981

QUEZEL, P. & S. Santa: Nouvelle Flore de l'Algérie et des régions désertiques meridionales, 2 Vols. Paris 1962–1963

SCHÖNFELDER, I. & P.: Kosmos-Atlas Mittelmeer- und Kanarenflora. 3. Stuttgart 2011

SCHÖNFELDER, P. & I.: Die neue Kosmos-Mittelmeerflora. Stuttgart 2008

SCHÖNFELDER, P. & I.: Die Kosmos-Kanaren-flora. 3. Stuttgart 2012

STRASSER, W.: Die Pflanzen der östlichen Ägäis inklusive Kreta und Zypern. Ruggell 2006

STRASSER, W.: Pflanzen des Peloponnes, 2. Ruggell 2002

STRID, A. & K. Tan: Flora Hellenica Band 1, 2. Königstein, Ruggell 1997, 2002

TUTIN, T. G. & al. (Hrsg.): Flora Europaea. 5 Vols, Cambridge 1964–1980, 1993

VALDÉS, B. & al.: Flora Vascular de Andalucía Occidental, 3 Vols. Barcelona 1987

VALDÉS, B. et al.: Catalogue des plantes vasculaires du Nord du Maroc, 2 Vols. Madrid 2002

ZOHARY, M.: Flora Palaestina. 4 text vols, 4 vols of plates. Jerusalem 1966–1986

Internet:

http://ww2.bgbm.org/EuroPlusMed/query.asp (Euro+Med Plantbase)

http://ww2.bgbm.org/mcl/query.asp (Med-Checklist)

http://luirig.altervista.org/flora/taxa/floraindice.php (Flora Italiana)

http://www.anthos.es/index.php?lang=en (Flora von Spanien)

http://flora.huji.ac.il/ (Flora of Israel)

Index

F indicates a species illustrated on the cover flaps

First published in the United Kingdom in 2017 by John Beaufoy Publishing Limited
11 Blenheim Court, 316 Woodstock Road, Oxford OX2 7NS, England

10 9 8 7 6 5 4 3 2 1

ISBN 978-1-912081-70-7

Translation: Lynda Hepburn
Editor, German edition: Claudia Salata
Design, German edition: eStudio Calamar

Printed and bound in Malaysia by Times Offset (M) Sdn Bhd

Illustration credits
D. Longo (875 photos), A. Scharlau (277), Wikimedia Commons H. Hillewaert (242) and P. Schönfelder 872 photos.

286 distribution maps P. Schönfelder, base map Wolfgang Lang.
109 botanical illustrations Marianne Golte-Bechtle (86) und Wolfgang Lang (23).

Pages 2 & 3: Giant Fennel, *Ferula communis*, Crete,
Page 4: Crown Daisy, *Glebionis coronaria*, Sicily,
Page 5: Poppy Anemone, *Anemone coronaria*, Turkey.

Cover design: eStudio Calamar with four photos by Peter Schönfelder.
Front cover: Field Daisy, *Pallenis maritime*, Bolonia, Spain
Back cover left to right, Crown Daisy *Glebionis coronaria*, Tricolour Convolvulus, *Convolvulus tricolor* and Cretan Cistus *Cistus creticus*

Red flowers

24

Pink Silk Tree
Albizia julibrissin

25

Mountain Ebony
Bauhinia variegata

26

Bottlebrush 'Captain Cook'
Callistemon viminalis

27

Trumpet Vine
Campsis radicans

28

Madagascar Periwinkle
Catharanthus roseu

29

Silk Floss Tree
Ceiba speciosa